LAUNCHING EUROPE

LAUNCHING EUROPE

AN ETHNOGRAPHY OF EUROPEAN COOPERATION IN SPACE SCIENCE

Stacia E. Zabusky

PRINCETON UNIVERSITY PRESS   PRINCETON, NEW JERSEY

Published by Princeton University Press, 41 William Street,
Princeton, New Jersey 08540
In the United Kingdom: Princeton University Press,
Chichester, West Sussex

Library of Congress Cataloging-in-Publication Data
Zabusky, Stacia E., 1959–
Launching Europe : an ethnography of European cooperation in space science / Stacia E. Zabusky.
p. cm.
Includes bibliographical references and index.
ISBN 0-691-03370-6. —ISBN 0-691-02972-5 (pbk.)
1. Astronauts—Europe—International cooperation. I. Title.
TL788.4.Z39 1995
629.4'094—dc20 94-32074 CIP

This book has been composed in Palatino

Princeton University Press books are printed on
acid-free paper and meet the guidelines
for permanence and durability of the Committee
on Production Guidelines for Book Longevity
of the Council on Library Resources

Printed in the United States of America

10 9 8 7 6 5 4 3 2 1

10 9 8 7 6 5 4 3 2 1

Contents

Tables

Preface

THINGS ARE changing fast in Europe these days. When I began the fieldwork on which this book is based, just over six years ago, the Berlin Wall was still firmly in place, Gorbachev was still pursuing glasnost and perestroika, and Western Europeans were approaching "Europe 1992" with mixed emotions, but feeling largely indifferent to the project of European integration. How different things are now. Still, the European Space Agency, site of my fieldwork, continues to produce space satellites, and participating scientists, engineers, technicians, and administrators continue to work together, coordinating, developing, and implementing space missions. It is these ongoing practices of mission development which are the focus of my ethnographic inquiries and which I analyze in the pages that follow; these practices continue even as the parameters of the work may have shifted in response to political, economic, and cultural developments in other arenas.

This combination of past context and ongoing venture presented me with a practical question as I was writing this book—whether or not to make use of the "ethnographic present." Finally, I chose to combine past and present tenses. I use the past tense to convey ethnographic details: specific events or observations, what people actually said to me or to each other, organizational or technical details that I know have changed. I use the present tense in two ways: in the description of generic aspects of structure (e.g., organizational features, mission technology, scientific findings) that continue to obtain; and in the discussion of analysis and interpretation of the data, since I am convinced that these continue to be accurate in relation to the specific set of data I collected.

Things are also changing fast in anthropology, in these as in other days. When I began fieldwork at ESA, few anthropologists were studying bureaucracy, science, and technology (although these were already objects of inquiry of whole subfields of sociology), and few anthropologists were studying the phenomenon of European integration. Again, how different things are now. While I had once envisioned the need to justify and defend, to an anthropological audience, this choice of inquiry, now books, articles, and conference presentations on precisely these topics appear at an amazing rate, suggesting wider acceptance and active pursuit of issues that only recently remained on the margins of anthropological consideration.

Nonetheless, I want to address, for a moment, the diverse kinds of readers I imagine coming to this text. The ethnography itself concerns how a group of people puts together technology to produce space satellites for scientific use. But this story is about much more than computer chips and clean rooms; it is about how these apparently passive and bloodless technological entities are enlivened in the practices of cooperation. They do not simply clutter up the landscape of the social process that is the focus of analysis; they play a key role in this process, as the medium through which people construct social, political, and existential connections as well as technical ones.

For those readers who come with a specific interest in the science and technology, for participants in ESA space science missions in particular, my depiction of the process will not necessarily fit your sense of the controlling power of technology in your efforts to produce space satellites. Here, the technology itself may appear to you to play a subordinate role, without the complexity that you engage in your work every day. I have presented the technical details—details of hardware, software, systems, even organizations, budgets, and politics—in this, perhaps simplified, way not only because of limited space and even limited understanding, but because the story I want to tell here, the story you have told me in your practice, emerges more clearly when the trees are thinned so that we can see the shape of the forest. Your job is to work on the trees, a job you execute admirably, but there is no less forest for all that. I recognize, however, that my pruning might not match your pruning—we see, most likely, quite different forests in the first place.

For those readers who come with an interest in cooperation, European integration, or social practices more generally, for anthropologists in particular, my depiction of the process may appear instead as boringly cluttered with technical minutiae. Such details may even deter you—I urge you, however, not to let the presence of technological components lull you into thinking that there is nothing here but dry details of no relevance to the larger issues anthropologists worry about. To do so is to be seduced by the very ideology of rationality that, as so many have warned us, masks new forms of domination. Instead, I invite you to look with me at how meaningful the technical is, and so to allow these participants, as stripped down and boring as they and their work may sometimes seem, to have passions that speak to you, and to us, passions that lead not only to the production of satellites but to dreams of the sacred as well.

Acknowledgments

As is the case with all ethnographies, the first debt of gratitude goes to those among whom we conduct our anthropological inquiries. I extend my thanks to all the staff at the Space Science Department at ESTEC for putting me up and for putting up with me during the nearly twelve months of my field research. Scientists, engineers, technicians, and secretaries all took time to talk to me, let me eavesdrop on their conversations, invited me to participate at times, and openly shared with me their thoughts, convictions, and feelings about the work that engaged them and eventually engaged me as well. In the course of my time at SSD, I also came to know many ESA staff members in other departments, as well as space science mission participants (for instance, members of Science Teams or instrument consortia) who were residents of other institutes and organizations. These people had even less reason to welcome me into their meetings, their offices, and their conversations, but they did so nonetheless; their insights significantly clarified and deepened my understanding of the complex process of working together on mission development.

This book began its life as a dissertation; indispensable in bringing that first manuscript to fruition were the members of my dissertation committee at Cornell University: Carol Greenhouse (now at Indiana University), Davydd Greenwood, and David Holmberg. Each of them contributed significantly to the genesis and evolution of the ethnographic and theoretical inquiries that I pursued in the thesis, and that continue to preoccupy me in this book. Carol Greenhouse was also a patient and generous reader as I worked on the book manuscript, in the process continuing to engage, support, and inspire me in myriad ways. Special thanks also go to Nancy Ries who has long served as mentor, friend, and debating partner. She carefully read most of the book revisions, offering both enthusiastic and critical comments, a combination that helped me to produce (I hope) better writing and better thinking. The faults that remain are, of course, entirely mine.

Many others, at Cornell and elsewhere, also provided both practical support and intellectual stimulation in the dissertation and book writing process. At ESTEC, I extend my thanks to Clare Bingham, Anne Cijsouw, Ann van den Eijkel, and Leonie Ouwerkerk; in different ways, they all helped make this book possible, whether by providing good cheer and company to me at ESTEC, or by helping me locate needed information about ESA. Thanks go to the two anthropologists who reviewed my manuscript for Princeton University Press, Michael Herzfeld and David Hess; both provided thoughtful critique and helpful suggestions. Editors Mary Murrell and Lauren Lepow at Princeton were always attentive, responsive, and full of goodwill, making the experience of publishing this manuscript a pleasure. And of course I extend my thanks to all of those who have engaged me in discussion and conversation on those anthropological issues which I pursue in this book: Steve Barley, John Borneman, Paul Donato, Jane Fajans, Toni Flores, Diana Forsythe, Martin Harwit, Bernd Lambert, John Norvell, Judy Rosenthal, Steve Sangren, Stefan Senders, John Weiss, and Jan Zeserson. Finally, I thank the Netherlands America Commission for Educational Exchange (through a Fulbright grant), the Council for European Studies, and both the Center for International Studies and the Graduate School at Cornell University for funding support at various stages of this project.

I also extend deep thanks to my family without whose support, love, and encouragement I could not have undertaken and persevered in a project of this magnitude. My parents, Charlotte and Norman Zabusky, have long fostered my intellectual and creative efforts, and in so doing have given me the confidence necessary to succeed. My brother Alexander and my sister Erica have supported me in this long endeavor, too, supplying sympathetic ears over many long-distance telephone calls during the years that this book was in the making. My deepest debt of gratitude and thanks goes to my husband, Donald Spector, whose patience and humor know no bounds. Don has learned to become an anthropologist over these many years; he participated in endless conversations about my analysis, read proposals and papers, offered critique and comfort, and shared with me a year in the Netherlands, when he sometimes helped me type up my field notes after several exhausting days of data collection.

Finally, I thank my children, Antal and Elias, both of whom have provided me with smiles and laughter that have sweetened this work more than I can say.

Abbreviations

AU	accounting unit
CERN	European Organisation for Nuclear Research
CO-I	coinvestigator
EC	European Community
ELDO	European Launcher Development Organisation
ESA	European Space Agency
ESOC	European Space Operations Centre
ESRIN	European Space Research Institute
ESRO	European Space Research Organisation
ESTEC	European Space Research and Technology Centre
PI	principal investigator
SSD	Space Science Department

LAUNCHING EUROPE

Introduction

Multiple Cooperations

> Cooperation within Europe—a patchwork of independent and autonomous states—is no simple matter, as so many experiences have shown. ESA's achievements are a reminder that successful cooperation can be achieved even in an area of growing economic importance. In this sense, space is making a contribution to European unity. . . . Europe as an entity will not be achieved overnight and not all at once. It will gradually come to exist as a result of practical achievements which in turn give rise to real solidarity.
>
> *(Reimar Lüst (1987), director general of ESA)*

> Better than 100,000 Europeans going [to other countries] on vacation is 1,000 working together on a project, because when you're a tourist, you're not committed to making compromises there, but if you have to get things to work, then you are forced to work together.
>
> *(An ESTEC project engineer)*

THIS BOOK is about cooperation. Substantively, it is about cooperation in Europe, as well as about cooperation in science and technology. Analytically, it is about cooperation as a form of structure and as a kind of practice. I explore these multiple cooperations—their intersections, antagonisms, and resonances—ethnographically, as they are manifest in the process of space science mission development at the European Space Agency (ESA).

Those participating in space science mission development are involved in cooperation on a grand scale. The organization that is ESA represents the joint effort of European governments, industries, and universities on a wide range of space activities, including, but by no means limited to, scientific research. This effort is part of a more encompassing drive to integrate the nation-states of Western Europe into a unified, or at least federated, entity, a European Community. It is also part of the tendency in industrialized nations to undertake the large-scale projects characteristic of "big science."

These structural factors are directly implicated in the daily work of scientists and engineers—in the most ordinary decision, the turn of a screw, a lunchtime conversation. Even on the local level, participants' work appears to be as much about cooperation as it is about technology. In other words, cooperation is not a matter of concern just to Eurocrats or technocrats; it is entangled in the very possibility, even the idea itself, of production, in this case, the production of scientific satellites to be launched into space. Images of successful launches and satellite operations are, indeed, omnipresent in daily activities, images such as the following:

> An Ariane rocket stands ready for launch. Its long, sleek, white exterior is painted with the logo of the European Space Agency, the flags of the thirteen member states, and the logo of the French space agency. Under the fairing at the nose of the rocket is a scientific satellite, also ready for launch. It is bulky, oddly shaped with long and short protuberances waiting to be extended once in orbit. It, too, is painted with the ESA logo and the flags of the thirteen member states. On launch, the rocket disappears from sight, tracked only by radar, visible to the Operations Team on the ground in light on a screen. Now the fairing opens and the satellite is boosted into orbit above the earth, unseen by human eyes. On the ground, people watch their consoles for the electromagnetic blips of information which signal to them that the satellite has survived and has begun its complex task of looking in space where human beings cannot look. (From field notes)

How do we get here, to this point of remarkable technology produced by remarkable teamwork, to this glorious launch and the glorification of European cooperation? I seek to answer this question here by exploring the practice of cooperation. Such cooperation is no mere fantasy; it is manifest concretely in ESA, one of the most successful European regional organizations. For thirty years, it has continued to produce, in the face of inadequate funding, superpower posturing, and shifts in European political priorities, science missions of the highest caliber (Dickson 1987). This study of science missions at ESA, an organization that has offered Western Europe a number of significant scientific and technological successes, thus offers an opportunity for understanding how cooperation can be achieved even in political, economic, and social contexts not entirely favorable to collaborative undertakings.

Contexts of Cooperation

ESA was established at a time when the movement toward European integration was beginning to flourish. This "regional impulse" (K. Twitchett 1980: 7) in Europe emerged in the aftermath of the devastation of World War II, when Western European political and economic leaders began to undertake joint ventures in an effort to "put an end to Europe's sorry history of conflict, bloodshed, suffering and destruction" (Borchardt 1987: 19). By developing such institutions, European leaders were expressing a variety of hopes. They hoped (and continue to hope) that integration would lead to a shared identity which would enable European citizens to overcome the antagonisms of nationalism (Bull 1993); they hoped that integration would enable each nation-state to derive greater economic benefits than any single one could achieve on its own; and they hoped that by pooling their economic and political resources they would be able to negotiate with the superpowers, particularly the United States, on equal terms, and so regain their stature in the international arena (Brucan 1988, Laqueur 1982, K. Twitchett 1980, Townshend 1980). These efforts found their most comprehensive expression in the establishment of the European Economic Community (founded in 1958), the supranational body that now dominates the regional scene. It is the European Community (EC) that, indeed, conjures up for most people the image of a uniting, if not united, Europe.[1]

In this historical context, cooperation has become the sine qua non of a contemporary "European" approach to economics, politics, and even society (see, for example, Borchardt 1987), so much so that integration is often asserted as a value in itself. In this way, interstate cooperation takes on a moral imperative—it will be the savior of Europe, delivering the continent from war, from poverty, and from backwardness. In this ideology, cooperation appears as the embodiment of peace, symbolizing the transcendence of the incessant warfare that has characterized European international relations for centuries. It is indicative of Europeans' "desire for peace" and of their "desire for a better, freer, juster world in which social and international relations would be conducted in a more orderly way" (Borchardt 1987: 6).

This is not to say that the drive toward integration is unanimously and constantly supported. From various vantage points and at certain moments cooperation can appear not as transcendence but as power, particularly the exercise of (supra)state power that threatens participants with the eradication of their heterogeneous identities in a homogenizing unity. For this reason, integration is often actively resisted by nation-states and by intrastate regional and ethnic populations, even as it is constructed by these same participants. In these countervailing forces, integration appears as a negative value, and interstate cooperation is contaminated by accusations of power mongering and greed. These centrifugal tendencies, toward, for instance, national sovereignty and local autonomy, have acted as counterweights to the centripetal forces unleashed after World War II, limiting their force and speed, but never quite changing the direction of what was happening.[2] The forces for integration continue apace, slowed but never stopped.

Integration is a concern not only of governments and populations; forces for integration have also characterized research and development in science and technology in the postwar era. As scientific ambitions have increased, the technology for scientific experimentation has grown correspondingly larger, more sophisticated, and more complex, complicating also the division of labor, as the participation of specialists representing a wide variety of disciplines and institutions is required to make such projects work. The massive scale of such projects turns attention to the importance of integration, both practically (getting all these people to work together) and technically (getting all the components to work together).

Big science also comes with a big budget, and those engaged in these large-scale activities turn to governments and industry to help them meet their technological and scientific goals. In this way, big science depends not only on practical and technical integration, but on political and economic cooperation as well to secure the considerable funds necessary to support such undertakings. In Europe, scientists have capitalized on the drive toward political cooperation, calling on Western European nation-states to support their projects in the interests of peace, prosperity, and the pursuit of knowledge. Organizations such as the European Organisation for Nuclear Research (better known as CERN), the European Molecular Biology Laboratory, the European Southern Observatory, the Joint European Torus, and of course the European Space Agency are some of the more prominent examples of the resulting entanglement of big science and European integration.[3]

In multiple domains, then, cooperation is viewed as a significant factor in the achievement of political-economic and practical-technical goals. Europeans at times almost sanctify cooperation as a social force, as I suggested above, seeing in it the only hope for a future of peace; cooperation appears as the antithesis of war, promising interdependent and mutually supporting, rather than mutually annihilating, differences. But achieving this kind of cooperation in the political sphere often seems to take more will than the nation-states of Europe can sustain. For this reason, political and industrial leaders, as well as scientists, often look to cooperation in the spheres of science and technology to lead them to the social outcome they desire; they hope that in the rational and orderly activities of science and technology, people (and states with them) will become interdependent almost without thinking about it, in their collective focus on mundane and technical details instead of the enmities, prejudices, and ambitions of politics.[4]

Cooperation Ethnographically

As suggested above, a multilayered cooperation structures and contextualizes the practices of participants in ESA space science missions. Yet in the course of my field research at ESTEC (the European Space Research and Technology Centre, ESA's primary production site), I quickly discovered that for the scientists and engineers involved with these missions, "cooperation" was a topic bedeviled by confusion and fraught with tension. I learned about such ambivalences particularly when I sat in at the margins of meetings, when the scientists, engineers, and technicians who worked regularly at ESTEC were joined by those participants who were based at external industries, institutes, and universities. Much of my ethnographic data indeed derive from my presence at these meetings as an observer.

Although the ESTEC staff quickly became accustomed to my presence in their midst (not only at such meetings, but also in the offices, halls, and cafeteria of the organization), at mission meetings with non-ESTEC participants I nonetheless found that I was regularly asked to introduce myself, and to announce the subject of my research for the benefit of those participants who had not yet met me or heard about my work. My standard response to these requests was to introduce myself as an anthropologist studying European cooperation in space science. The reaction to this was, invariably, laughter, followed by wry comments—"If you see any examples of people cooperating, let us know";"There's none of that here"—and an exchange of knowing glances. Although I initially dismissed this response, as I heard it repeated many times I gradually came to realize that it was significant. The message was that people's primary experience was not one of cooperation, by which (as I later understood) they took me to mean agreement, harmony, and unity. If indeed that was what I sought, these participants insisted that I could not find it there, in the heart of their process, where working things out often entailed fighting things out. Their primary experience was instead one of conflict, of contrary opinions and competing interests; in their view, all of this was the antithesis of something that they might call cooperation.

Yet in their laughter they did not mean to dismiss my project as impossible. Rather, they seemed to be expressing a hope that I might be able to tell them what was wrong with how they did work together, why there was so much conflict, and how they could overcome this in order to interact in a harmonious spirit. They expressed this hope, for the most part, indirectly. For instance, during meetings, I would sit at the back of the room, scribbling in my notebook, paying careful attention to the proceedings, but keeping my head down, trying to keep myself apart from the action. The participants generally ignored me, but at certain key moments, moments of confrontation, they would become explicitly aware of my presence. Their awareness of me manifested itself in various ways. In the midst of or following a heated exchange, someone might turn to me and say, "Are you getting this all down?"; alternatively, someone might refer to me, saying, for instance, jokingly, "Watch out, Stacia is listening in here." Others would take a less obtrusive tack, waiting until the coffee break to comment to me, "That must have been particularly interesting for you." By becoming aware of my presence at these moments of conflict and tension, the participants highlighted for me that these were the troublesome (if exciting) interactions which, from their perspective, I must endeavor to explain.

But this was not the whole story. Despite the initial response of laughter and the deprecating comments about the lack of cooperation here and now, participants' commentary for my benefit often made the opposite case. During the first break in the meeting, people would come up to me to say, "Well, of course we cooperate"; as proof they adduced the successful operation of ESA or the success of scientific missions. Sometimes, in thus "illustrating" the achievement of cooperation, people argued that it was the complexity of the science and the technology which mandated such cooperation (in the logic of "big science"). Others argued that it was simply political and economic expediency that brought them together. But there were also many who grew passionate in their defense of a cooperation that they saw as having moral implications which transcended technical benefits and cost effectiveness and indeed expressed the soul of European culture.

Participants thus both denied their participation in cooperation and simultaneously gave me proof that cooperation existed, leaving me with an essential paradox, located at the heart of their practices, which required explanation. What did people mean when they defended, denied, or more generally talked about "cooperation"? What were the connections among the varied meanings of cooperation (and the tensions accompanying these different meanings), the structure of ESA—an organization in which people worked to produce such concrete outcomes as Ariane rockets and scientific satellites—and the wider political context of European integration? And finally, why were the participants' reactions to my project so emotional, as they expressed both anger and hope, both cynicism and frustration? In other words, how could I make sense of the connections between the structure of cooperation and individual participants' experiences in their daily work, a kind of work that appeared to offer not only intellectual but existential challenges to those involved? These are the ethnographic questions that organize this book, as I consider cooperation as structure, practice, and aspiration.

Cooperation Analytically

In my focus on cooperation, I take up anthropology's classic attention to issues concerning the division of labor and social cooperation. Like many anthropologists, I am interested in the way in which people in a heterogeneous social, cultural, and political environment collectively produce a social system that has, if not unity, then coherence. The heterogeneity to which I refer here corresponds primarily to differentiation according to the tasks and goals of production, a differentiation that is part of the division of labor in social systems generally speaking. It is in terms of such differentiation that cooperation emerges as a meaningful form of social interaction, political action, and moral suasion, as Durkheim (1933) taught us long ago. The questions about cooperation that I outlined above ask how such differentiation is perceived and experienced by participants in cooperation. In other words, these questions are about the meanings and practices of the division of labor that structures the development of ESA space science missions.

My analysis of work on such missions takes into account certain observations made by scholars of organizations: inside any organization the bureaucratic structures that define positions and tasks do not themselves describe the actions and interactions of those who work in terms of their rules and regulations. Rather, bureaucratic hierarchies of status and function serve only to delimit the scope of action without necessarily eliminating people's ability to improvise in their spaces; moreover, the rationalized structures of bureaucracies do not succeed in eradicating the play of meaning from within their spaces, even if that is, in part, their goal.[5] This means that I treat ESA here not simply as an institution that organizes and funds space missions and the people who work on them, but as a context that both structures and is structured by the social practices and cultural values of those who carry out the scientific and technical work of mission development.

More generally, I consider in my analysis the dialectical relationship between the structural matrix of the division of labor, as it is articulated by organization and in technology, and practical improvisation, as undertaken by individuals negotiating such structural edifices. To do so, I offer a "thick description" (Geertz 1973) of this practice of cooperation, attending to the different ways in which participants perceive such structures (and negotiate these perceptions). My invocation of Geertz here is intended primarily to signal the fundamentally cultural orientation of my analysis, with its concern for explicating the contested meanings that accompany the practices at issue here. I do not, however, attempt to delineate the neat contours of some abstract "cultural pattern," situated above practical realities and somehow controlling them. Instead, I attempt to show how social actors themselves manipulate and contest such "regulating" patterns through "tactics" that "do not obey the law of the place, for they are not defined or identified by it" (Certeau 1984: 29). Indeed, I will show that the practice of cooperation depends on the ongoing negotiation of the (often irreconcilable) differences put into play by the division of labor.

Despite the negotiations of ambiguity characterizing their quotidian activities, however, participants themselves appear to hold to a Durkheimian view of cooperation and a rational division of labor. That is, they expect to find the affective consensus of "organic solidarity" or even of "effervescence" when engaged in the cooperative tasks of mission development. In part, such expectations reflect the powerful presence of the ideologies of both European integration and big science in the daily work of science and technology. These ideologies, as was evident above, themselves reproduce a rather Durkheimian vision of cooperation as "social integration": they emphasize that the division of labor (whether political-economic or practical-technical) is based on an organization of difference which permits rational and solidary relationships on every level. In fact, these ideologies also assert that without the cooperation which such organization demands, individual participants (whether nations, instruments, or people) cannot realize their full potential.

This same vision, however, leads to participants' simultaneous denials of cooperation. These denials arise partly from their (disappointed) recognition that their local practices do not correspond to the rational and solidary order of "cooperation." In *practice*, participants take the diversity embedded in the division of labor and transform it from a principle that *organizes* structure into a cultural value that *disorganizes* temporal activities and keeps everything moving—and everybody arguing. At times, then, when they deny cooperation, they are articulating their frustration that instead of the harmony and unity promised by participation in cooperative efforts, they find social unraveling, cultural upheaval, and the constant threat of anomie in their daily routines.

But these denials are not only expressions of passive cynicism; they are equally expressions of active resistance to the very vision of cooperation that participants, at other times, seem to share. The disorder of diversity that motivates practice is, in fact, valued by participants, and by denying cooperation, participants refuse to be implicated in those hegemonic assertions of unity which ideologies of cooperation proclaim. These ideologies, because they stress the achievement of rational solidarity, discount the reality of those everyday practices in which participants find such "pleasure in getting around the rules of a constraining space" (Certeau 1984: 18). The ideology of rational solidarity indeed by definition silences the very conflict on which cooperation is built and on which it depends, thus rendering disorder invisible.

When participants deny cooperation, they are rejecting, among other things, the way that it legitimates domination. At these moments, they resist cooperation because they recognize in its elimination of diversity the "strategy" (ibid.: xix) of an order that seeks to confine them, and in its silencing of conflict the silencing of their own voices, clamoring to be heard in the din of practice. Nonetheless, there are also moments when participants affirm cooperation and regret its absence. When they do this, they are articulating their perception that cooperation may grant them connection to peers and colleagues, a kind of connection that is both instrumental and existential, that makes possible not only the production of artifacts but their own individuality as well.

The presence of these contending cultural forms of cooperation accounts for participants' complex and ambivalent responses to my statements about the subject of my research. In fact, when they referred to their everyday work, they did not use the word *cooperation*, which I had assumed described the process in which they were engaged, and which anthropologists might use analytically to describe the way in which people practically integrate their productive activities based in a division of labor. Participants did not, in fact, have a name for the process in which they were involved on a day-to-day basis; they were certainly not "cooperating," since in their view, as I have suggested above, that would involve harmony, order, and silence. What they *were* doing was, simply, "working together"; this system required no name because it was the very thing that people inhabited as they did their work.

Working together is a process that depends on disorder and ambiguity, on independence and interest confronting and circling about each other in noisy altercations, the perpetual conflicts of diversity. This is the negotiation of the differences of the division of labor, negotiations that characterize the practice of cooperation; my exploration of this practice forms the substance of this book's central chapters.

A Brief Comment on Unity

Although I have emphasized here the way in which the practice of cooperation depends on negotiation, ambiguity, and conflict, I do not intend thereby to exclude from the domain of practice the possibility and the achievement of solidarity altogether; to do so would be to suggest, as it were, that there is no cooperation in cooperation. In fact, I believe that people do, from time to time, experience moments of pure connection when they are working together, and that it is such moments, as I argue in the Epilogue, which motivate participants to keep enduring the structures of cooperation from which they distinguish their practice. These occasions of connection are mere flashes, occupying only an instant—they are epiphanies, when people engaged in action are able to focus on what is before them, when they *concentrate* (to use a term borrowed from Simmel [1955]). In this concentrating together, people aspire to transcend the distractions and distinctions that divide them in the practical and political world of cooperation. In such moments, they endeavor to connect, to discover, and to create; in so doing, they are liberated both from oppressive unities and from the anomic pressures of differentiation.

I point out here the possibility of connection because this book's central argument is that the promise of unity made in the rhetoric and the structure of cooperation represents an ideological force which does not correspond either to the experience of or values in local practices. Participants in fact resist the hegemony of cooperation by asserting the anarchic importance of diversity and difference and the constant negotiation these require. In taking seriously these expressions of resistance, I follow the lead of many contemporary anthropologists (e.g., Clifford and Marcus 1986, Herzfeld 1987, Marcus and Fischer 1986, Taussig 1987) who advocate the inclusion of informants' voices in ethnographic texts, in part as an exercise in critical reflection on the enterprise of ethnography itself. This reflection is particularly necessary, such critics argue, because anthropology's very search for social order in local venues itself represents the imposition of Western social scientific concerns with order on ordinary people's lives, lives that engage in chaotic and ambiguous (if generative) practices rather than submit to neat and orderly routines. In other words, instead of organization, anthropologists should pay attention to disorganization; instead of consensus, we should listen for "dissensus."

Herzfeld (1987: 46) indeed argues that anthropology has been in the thrall of a European "statist ideology" which "emphasizes conformity, unity and perfection" and which accordingly "tries to banish the knowledge of a past fall behind the present perfections of a created order. Ordinary people, however, know better: they understand the social character of self-interest, and they know that it affects the highest levels of political activity" (ibid.: 153–54). In Herzfeld's view, the scholarly search for cultural consensus, for the "seamless web" of cultural meaning, reflects not the existence of such a consensus in "ordinary life" but rather the scholar's own involvement in the elite project of state building and state protecting. Taussig (1987: 441) is even more adamant on this score; daily life is characterized by "alterations, cracks, displacements, and sudden interruptions," and not by the order and unity that so many anthropologists have found in social systems everywhere. Looking at ritual from his perspective, for instance, Taussig finds not a mechanism that insures social cohesion by providing the (temporary) experience of *communitas* (Victor Turner's [1974] version of Durkheimian solidarity), but instead "awkwardness of fit, breaking-up and scrambling, the predominance of the left hand and of anarchy" (Taussig 1987: 442).

I would argue, however, that it is too easy to dismiss all desire for connection, and the order that may imply, as simply a reflection of the intrusion and imposition of power into local experiences. Drawing a clear distinction between the chaotic lives of "ordinary people" and the regimentation of states and the elites who support them itself polarizes what cannot be so cleanly divided. "Ordinary people" may recognize in their daily lives processes that are rife with conflict and fraught with interest and division, and they may even assert these in resistance to the organizing and unifying forces of state regimes, colonial laws, and bureaucratic regulations. But this is not the only experience of and in practice. Habermas (1989a: 154) indeed argues that in the ordinary, everyday practical contexts of "communicative action," participants interact "expressly for the cooperatively pursued goal of reaching understanding." In such moments, "it is the speakers and hearers themselves who seek consensus and measure it against truth, rightness, and sincerity." Moreover, to insist that all desires for unity are the result of exercises of political domination in effect denies "ordinary people" their own capacity to dream.

Outline of the Book

The chapters of this book move from structure through practice and back to structure again. In the course of this linear story, I endeavor to convey the multiplex dimensions of cooperation by providing a circulating, layered narrative, in which the same issues and problems repeatedly surface in the consideration of different aspects of the problem. What in practice is messy, involuted, and mutually implicating thus appears here as somewhat forward-moving, indeed straightforward and linear. Yet this narrative's sense of direction, a direction that leads the reader toward a culminating production (of artifacts and structures) is not simply one which I have imposed on this text. This push toward production comes from the practice itself; the linearity and the closure are part of the story the participants themselves tell about their practice. Although on one level, then, this narrative has the neat and satisfying orderliness of structure, the story of working together articulated with this order is—as I hope readers will discover—one of dynamism, disorder, and dialectic. It does not come to rest, even if things are produced. Indeed, it is in part for this reason that I do not let the narrative rest in chapter 7, artifacts produced, all said and done, but push onward to the Epilogue, where I take the analysis just completed and complicate it once more in dreams of the sacred.

The first three chapters of the book provide an overview of theoretical issues related to cooperation (chapter 1), the historical and structural context of ESA and its missions (chapter 2), and the local parameters of working together, including all the myriad details of where, who, and what working together involves (chapter 3). Chapters 4, 5, and 6 take up the analysis of working together itself, layer upon layer: social and cultural aspects (chapter 4), political aspects (chapter 5), and existential aspects (chapter 6), separated analytically, but entangled in practice. Chapter 7 returns to structure once more and explores its relation to the practices just considered. Finally, the Epilogue locates the analysis of working together in a wider cultural context, as it takes up the dreams of modernity and offers a theology of cooperation.

Summary of the Analysis

In the course of working together, people negotiate the contrary terrain of diversity, the local expression of the conditions of the division of labor. This terrain involves participants in dynamic social practices that are the stuff of which working together is made: over and over again, participants now construct and now deconstruct specific social groups, as they revel in conflict and appeal to harmony; over and over again, they evade others' assertions of superiority and make their own, and in so doing, they take responsibility and shirk it; over and over again, they celebrate their work and idle away the time. Throughout these practices, the explicit focus is on the technical, on the hardware, the software, and on the scientific ideas that the technology is designed to execute and answer.

Working together is ultimately a productive process—people are explicitly trying to construct functioning technical artifacts. Missions do succeed (that is, satellites are launched, orbit or trace out a flight path, detect light signals or other information, and send back data for analysis), and in this tangible success, another, less obvious outcome is revealed as well—cooperation. Both technology and cooperation (in the form of ESA and in the form of successful missions) represent the persistence of enduring structures, themselves made possible because many individuals and independent groups somehow manage to figure out a way to sit down and get some work done together, despite the fact that not only is no one really in charge, no one knows entirely what is going on everywhere else. This is the ongoing puzzle of working together that, no matter how many instruments are made or satellites launched, is never definitively solved.

What this means is that even though working together is on one level a goal-directed process (to build this piece of hardware, to design this software system, to launch this satellite, and so on), on another level it remains indeterminate and fluid. When people produce things, they run the risk of becoming alienated from the products of their own labor; hence, they cannot stop there. Even as satellites fly and bureaucracies endure, therefore, people continue to immerse themselves in the ongoing struggle of working together in order to stay excited, and thus alive. In so doing, they may lose sight of their connection to the structures their practices have made possible, but at the same time they find freedom in this social system, a system constructed out of and continually disturbed by the negotiation of diversity.

It is in fact the effort of trying to solve this (technical and social) puzzle that keeps participants excited about and committed to their work. If the process stops, either because it solidifies in artifacts or stagnates in cooperative structures, there is a kind of death, and this is what participants most actively resist. At the heart of working together, then, is the necessity of keeping the process going, of not solving the puzzle, because it is in the struggle itself that people feel themselves to be alive.

This life-and-death struggle acquires particular urgency and poignancy, moreover, from its connection to the immanence of the sacred. In the excitements of working together, people not only undo the solidity of structures of cooperation but aspire to a kind of transcendent enlightenment. This story can be told, however, only after the production of artifacts is finished, and I return to it in the space of dreams that the Epilogue reveals.

One

The Study of Cooperation: Theoretical Issues

> At the peak of this mission's development, we will have more than one thousand people working together, most of whom I don't know, who have to be coordinated, and not run amiss of each other.
> *(A senior ESTEC engineer)*

> Basically cooperation works. If you enter into it openly, you actually learn things from having different traditions present, and seeing how they clash and how they supplement each other. Obviously they clash and there are difficulties, but you can still get something extra out of that interaction.
> *(An ESTEC project engineer)*

THE ANTHROPOLOGICAL approach to cooperation I undertake in this book stresses social and cultural dimensions rather than individual motivations or interests; it therefore includes considerations of context—social, cultural, political—as critical parts of analysis. Such emphases distinguish this study of cooperation from those undertaken by scholars in other disciplines, such as social psychology or decision making. The latter types of studies make use of experiment and computer modeling, rather than ethnography or textual analysis, to address, for instance, the social parameters that generate cooperative behavior, the psychological traits correlated with cooperativeness or helpfulness, or conditions under which self-interested individuals can maximize rewards by engaging in cooperation (see, for instance, Derlega and Grzelak 1982 and Axelrod 1984). Such "contrived or experimental situations" (Nisbet 1968: 389) that have dominated the study of cooperation, however, tell us little about what actually happens when a group of people engages in this kind of social process in the "real world." As a result, as one game theorist (Axelrod 1984: 190) himself candidly acknowledged, theories of cooperation built on such studies do not take into account "complicating" factors such as "ideology, bureaucratic politics, commitments, coalitions, mediation, and leadership," or even, as a psychologist put it, "the satisfaction of having achieved one's goals, whatever these goals might be" (Rabbie 1982: 139). Moreover, the experimental designs used in such studies tend to be predicated on economistic and individualistic models of human behavior (Barley 1991). The comparative and ethnographic orientation of anthropology challenges these largely unquestioned biases of psychological and decision-making approaches precisely because it focuses on the cultural contexts of action. Decisions are not made, and people do not cooperate, in environments devoid of complications or heterogeneous elements.

This is not to say that there is nothing to be learned from these other types of studies. But the questions that such studies often ask—"When does such activity occur?" or "How can we generate cooperative behavior?"—in the end treat cooperation as a black box, something that is readily identifiable and that can be compared to and contrasted with such other entities as "competition" or "conflict" in an abstract, even reified sense. In this book, I open up the black box by trying to understand not only how cooperation works, but what it means, to those who are involved in the process of "working together towards one end" as Mead (1961: 8) defined cooperation in her 1937 study *Cooperation and Competition among Primitive Peoples.*

In the decades since Mead's volume was published, there has been a decided silence about or lack of interest in cooperation as a theoretical issue in anthropology; indeed, it only rarely appears as an entry in indexes to ethnographic texts. This reflects a tendency to conflate cooperation with social order or social integration, a tendency that ignores the existence of something I have called the practice of cooperation. Analysts have tended to assume cooperation, leaving it as a kind of unexamined, residual category, what obtains when people are not competing or fighting. Conflict appears, by contrast, as the more compelling object of inquiry; as Nader (1968: 237) argues, it is "more readily observable than is integration. As a result, much of anthropological discussion of integration . . . is implicit rather than explicit."

Indeed, anthropologists have devoted a great deal of attention to conflict (Howell and Willis 1989, Sluka 1992: 21; I consider the relationship between conflict and cooperation in more detail below). Howell and Willis (1989: 13) argue that this focus, in British social anthropology at least, reflects "the implicit assumption . . . that humans are violent, and that society in various ways controls and constrains." As a result, anthropologists have investigated these "various ways," seeking out and exploring the myriad forms and expressions of conflict and its resolution. Conflict seems to be everywhere; cooperation (or "peace"), if it appears at all, is just a temporary truce in an ongoing battle of aggressive impulses: "it is the absence of conflict, not the presence of something else that is noted" (ibid.: 8). Howell and Willis argue instead for something more positive, in a substantive rather than an ethical sense. They insist, drawing on work of Trevarthen and Logotheti (1989) and Carrithers (1989), that "the important point to stress is that humans appear to possess an innate capacity for finding common cause in a great number of cultural activities" (Howell and Willis 1989: 19); thus, they argue for a view which recognizes that "co-operation, not competition, is more conducive for survival" (ibid.: 21).

I focus on the concept of cooperation in this book not only because, as I argued in the Introduction, it is cooperation that, in various forms, contextualizes and drives the practices that interest me. I focus on cooperation in part because I am making an effort to define cooperation positively (although I am not making an evolutionary argument, as Howell and Willis are, at least in part) and not as an afterthought. I do so in order to bring out what has been left implicit in studies of conflict, not to deny the presence or significance of conflict—that would simply reverse the polarities. Instead, I demonstrate here that cooperation must be understood as evidence of survival in an existential sense; accordingly, it must be understood as something that is actively accomplished in terms of conflict and not despite or without it.

Specifically, I treat cooperation as a particular kind of collective action, one oriented toward a distinct goal, whatever its nature (material, social, economic, or other). I argue that cooperation is, in essence, a process of production. In so defining cooperation, I remain sensitive to the fact that cooperation can be understood in many ways, both by analysts and by those involved in the day-to-day activities of something we and they identify as cooperation. At the very least, as Nisbet (1968: 384) points out, it can be regarded "as an ethical norm, as a social process, or as an institutional structure." The question is, what is the relationship among these different understandings of cooperation? I answer this question ethnographically in the body of the book, by turning attention to questions such as these: What are those involved in such processes doing? How do they describe what they are doing? How does the nature of the goal or outcome itself affect the organization, dynamics, and meaning of the activity? In this chapter, I relate such ethnographic questions about cooperation to some key issues and topics in anthropological and sociological literatures: structure and practice, social order and conflict, power and hierarchy, Europe and diversity, and science and technology.

Cooperation, Structure, and Practice

I approach the study of cooperation from the perspective of "practice theory" (Ortner 1984). This means that I explore the dynamic, temporal processes of everyday work (the practice of "working together") and the relation this practice bears to the more orderly and organized structures that seem to surround participants at every turn (structures that participants recognize as "cooperation"). These structures, in Giddens's (1984: xxxi) words, provide the "rules and resources" for action. Such structures are not only contexts for action, however; Giddens argues that "the structural properties of social systems are both medium and outcome of the practices they discursively organize" (ibid.: 25). Structure is an intrinsic part of practice; furthermore, it is through their enactment of structural principles that people (re)produce those very structures which seem, at other moments, to exist independently of their actions (and appear, perhaps, to constrain or oppress them).

Structure is, thus, connected to and implicated in practice, although that connection is at times obscured for participants. Giddens (1984: 17) indeed insists that "structure exists . . . only in its instantiations in . . . practice" even though participants' "own theories . . . may reify those structures" (ibid.: 25). Although agents' theories—or, in Bourdieu's (1977) terms, their reflections and analytic understandings—may make concrete what is otherwise only immanent, their practice in fact keeps structure always provisional. And as I will show in the case at hand, participants on ESA space science missions prefer to keep structure provisional, because only thus, they believe, can they keep themselves free. Freedom is, indeed, a prerequisite—in the view of participants—both to successful working together and to successful outcomes (e.g., space satellites) of that process.

Participants undermine their own theories and reifications of cooperation by immersing themselves repeatedly in the "incoherent coherence" (Bourdieu 1977: 158) of practice, by, in short, "working together." This is the "practical consciousness" (Giddens 1984: xxiii) of cooperation. Giddens describes such consciousness as a kind of comprehension that "consists of all the things which actors know tacitly about how to 'go on' in the contexts of social life without being able to give them direct discursive expression." Bourdieu (1977) also makes such a distinction, noting that this awareness of how to "go on," which he calls "practical knowledge," may draw on analytic understandings but is not equivalent to them: "practice by definition excludes theoretical questions" (ibid.: 106), partly because "agents cannot afford the luxury of logical speculation, mystical effusions or metaphysical anxiety" (ibid.: 115).

In this view, there is a generative relationship between "practical knowledge" and symbolic systems (see also Piaget 1970 and Turner 1977 for related theories of generative and contextualizing structures). Temporality plays a significant role in this relationship. Whereas analytic knowledge is constituted outside of time and space, practical knowledge by definition can exist only *in* time and space. Whereas analytic knowledge takes the form of a totalizing order that rationalizes and organizes contradictions and irregularities, practice dissolves this order, scattering its pieces into the flow of time where they swirl about each other, crisscrossing and separating in agents' strategies of action. It is time that allows this unraveling of what otherwise seems so stable; for this reason, practical consciousness can sustain contingency, contradiction, and ambiguity where analytic reflection cannot. Bourdieu (ibid.: 105) illustrates this contrast by juxtaposing the structure of genealogy and the practice of kinship: "Genealogy substitutes a space of unequivocal, homogeneous relationships, established once and for all, for a spatially and temporally discontinuous set of islands of kinship, ranked and organized to suit the needs of the moment and brought into practical existence gradually and intermittently." Both kinds of knowledge are connected in a generative, dialectical process of transformation and accommodation. Practice is not, in other words, reducible to the rules of structure; neither is the order of structure rendered false by the disorder of practice. Each depends on the other, and the people who inhabit their social systems require both as well in order to survive.

Working together can be understood, from this perspective, as an enacted and even embodied knowledge rather than an abstract theory articulated in reflection. This is not to say that participants in working together do not know anything about what they do, or why they do it, but that their consciousness of their actions is expressed through action, and not in reflection. I do not mean to suggest, however, that this understanding of working together is never given discursive expression, as my ethnographic data make clear. Giddens (1984: 4) notes, after all, that "the line between discursive and practical consciousness is fluctuating and permeable, both in the experience of individual agents and as regards comparisons between actors in different contexts of social activity."

Those involved in working together *do* reflect on and talk about their practices, particularly when making comparisons—between, for instance, their own "egalitarian, cooperative" work practices in space science and what they perceived as the "command-and-obey" form of American work practices, particularly at NASA. In fact, my informants proved themselves to be expert theoreticians when I questioned them directly about such abstract concepts as "cooperation." During these moments, they would muse about the shape and structure of "cooperation," pondering its requirements, its politics, and its problematics (this is what Bourdieu [1977: 18] refers to as a "semi-theoretical" orientation). In the contexts of daily work, however, people did not sit around trying to figure out how to work together in relation to the structures and ideologies of cooperation; they simply *did* it, resolving problems in action, and in technical action quite specifically. This is the essence of "practical consciousness" or "practical knowledge." This is one reason why participants have no name for the practices in which they engage; it is only from the analytic perspective that something called "working together" can be recognized and described (I discuss this distinction as it appears ethnographically in chapter 3).

I have pursued this extended discussion of the relationship between structure and practice not just because these ideas are relevant to theories of cooperation, but because it reflects my theoretical orientation more generally.[1] In the approach I am undertaking here, the scholarly problem is less to describe and/or derive structural systems and principles that insure order than to account for the dialectical logic inherent in the very contradictions in behavior. Practice theory seeks to restore to local actors their agency even in constraining contexts; attention is thus drawn to the way individuals enact, reproduce, and resist social structures through the (largely but not entirely) unconscious embodiment of—and resistance to—the "doxa" (Bourdieu 1977) of consensual belief. Determining the structural shape of cooperation, then, does not solve the problem of local practice that may make use of, refer to, and even produce such a structure, but that through time appears much less organized. Indeed, the practice of cooperation, as I stated in the Introduction and as I show in the following chapters, consists of the ongoing negotiation of the often irreconcilable differences put into play by the division of labor, a negotiation that itself proceeds through conflict and ambiguity as much as through solidarity and orderliness.

Cooperation, Social Order, and Conflict

Understanding the division of labor is key to understanding the structure of cooperation both analytically and as it appears to participants on ESA space science missions. In his influential text, Durkheim (1933) analyzed the division of labor in industrial capitalism and proposed cooperation as the basis of social order in such societies. He argued that the highly specialized division of labor characteristic of industrial capitalism was making individuals ever more dependent on society.

This dependence came about because the specialization of tasks and functions inherent in the division of labor did not produce "a jumble of juxtaposed atoms" but instead established the basis for social cohesion, or "organic solidarity": "each of the functions that [the members of society] . . . exercise is, in a fixed way, dependent upon others, and with them forms a solidary system" (ibid.: 227). Cooperation is in this way not only a rational mode of organization accompanied by social as well as technical integration, but also one that has, simultaneously, an "intrinsic morality" (ibid.: 228). Indeed, when Durkheim argued that the division of labor promotes "solidarity," he was arguing that it had not only technical-functional dimensions but normative ones as well. In other words, as Giddens (1971: 79) explicates Durkheim, "contemporary society is still a moral order," even if it lacks "a strongly defined moral consensus." I pursue the implications of an ideal form of cooperation more fully in the Epilogue; here, I focus on issues of social order and their implications for conflict.[2]

Durkheim's discussion establishes a clear framework for understanding the structures of the division of labor, that "organization of relationships" (Giddens 1971: 68) which constitutes complex, capitalist society. This emphasis on, in Lukes's (1982: 21) words, "social integration as the proximate goal of enlightened political action" in systems of cooperation based on the division of labor draws attention to the importance of "conflict, power and unpredictability" (ibid.: 23) in society, although Durkheim himself did not explore these "disorderly" dimensions more fully. As I indicated above, my own approach pays significant attention precisely to such "disorderly" elements, in part because I examine cooperation not just as structure but as practice. By locating agents thus in the structures of the division of labor, I hope to animate those stark matrices of thought and action which Durkheim delineated.

Many anthropologists following in Durkheim's footsteps have been deeply interested in social integration as well, but they have been drawn to exploring just those issues which Durkheim de-emphasized—in particular, the issue of conflict. This makes sense, since in the intellectual tradition of Durkheimian sociology, conflict appears to pose a significant problem in the way that it threatens the presupposed social order. British structural-functionalists and their followers, for instance, were preoccupied with this problem; they theorized that conflict was part of, and subsumed in, the maintenance of social order, insuring equilibrium rather than threatening it (e.g., Evans-Pritchard 1940, Fortes 1959, Gluckman 1956, Turner 1968). A similar view also characterized the approaches of American anthropologists studying dispute resolution and legal systems (e.g., Bohannon 1967, Nader and

Todd 1978). For these analysts, conflict was typically viewed as a process that helped to constitute, and resolve contradictions in, social order; accordingly, the questions asked concerned how conflict was manifested and how it was resolved or contained in an overarching context of integration. But as Sluka (1992) points out in a review article, the intense interest in conflict in anthropology has not resulted in theoretical consensus on the issue. Indeed, although there have been numerous ethnographic considerations of conflict, "a high degree of theoretical eclecticism" characterizes the field, partly because "'conflict' is a broadly defined concept and widely interpreted phenomenon" (ibid.: 20).

In this book, one reason for my interest in conflict is that conflict was a significant issue for participants on European space science missions (see chapter 4). Participants, however, gave contradictory messages about conflict—at times it appeared as a negative value, a source of unproductive and anomic experiences; at other times it appeared as a positive value, a source of individuality and creativity. In this sense, conflict was not merely subordinate to integration, nor was it simply a threat to integration, nor was it more essential than integration—it was all of these things. In my understanding of working together, then, I return to the most basic of insights: "Conflict and cooperation are not separable things, but phases of one process which always involves something of both" (Charles Cooley quoted in Coser 1968: 236). The process in question here is one of production; in this process, both cooperation and conflict are equally significant, each calling the other into play and each constraining the complete development of the other. Cooperation as a productive process, then, manifests a dynamic counterpoint between social processes of integration and of dispersal; this is analogous to, but not identical with, a cultural dynamic of harmony and conflict, as I argue in chapter 4.

This classic kind of opposition in social forces was described by Durkheim (1915: 391) in his discussion of the "rhythm of . . . social life," not among the workers of Western capitalist society but among aboriginal tribes in Australia. Durkheim described a rhythm between integrating ritual events and dispersing daily activities, and he argued that both dimensions were a necessary part of social life:

> Society is able to revivify the sentiment it has of itself only by assembling. But it cannot be assembled all the time. The exigencies of life do not allow it to remain in congregation indefinitely; so it scatters, to assemble anew when it again feels the need of this. (Ibid.)

Such "exigencies of life" are found in the "utilitarian and individual avocations which take the greater part of the attention" on "ordinary days" (ibid.: 389–90)—these are profane concerns, in contrast to the sacred concerns of ritual and ceremony, wherein "the spark of a social being" is conjured up and given life by participants.

Later ethnographic studies by structural-functionalists built on Durkheim's analysis, as I suggested above; these, too, discovered a relationship between centrifugal and centripetal forces in social systems in other parts of the world, most particularly in African settings. For example, Evans-Pritchard (1940), Gluckman (1956), and Fortes (1959) all discussed the way in which social groups persist in a state of relative equilibrium by establishing the "peace in the feud"; that is, people create a balance between those forces which link them to one another and those which drive them apart in order to maintain their social orders. This structural-functionalist model has been justly criticized for its emphasis on the dubious achievement of equilibrium at the expense of the uncertainties of conflict, and its overemphasis on order at the expense of process and contradiction.

The fact that there is a relationship between the forces of "fission" and "fusion" cannot be discounted; however, the goal of recent discussions of this relationship is not to focus on the way such an interaction creates order and equilibrium. Instead, anthropologists discuss the ways in which agents experience and produce their social systems out of the movement and ambiguities inherent in the dialectical interaction of such forces. For example, both Myers (1986) and Holmberg (1989) provide analyses of social and ritual life in small-scale societies (Australian aborigines and Tamang in Nepal, respectively) in which there is a "rhythm" between aggregation and dispersion that is itself the substance of local practice and the source of its attendant values.

Gerlach and Radcliffe (1979) also recognize this kind of push me–pull you process at work in the increasingly interconnected world of the industrial West, suggesting that the processual dynamic of such counterbalancing forces is not confined to small-scale societies. In their analysis of conflicts that local citizens have with big companies and big government over energy and land use in Western (European and American) nation-states, Gerlach and Radcliffe describe development as a process that shifts between "independence" and "interdependence," which are the names they give to forces others describe as fission and fusion. In this constant shifting, with both values enunciated in both local and national-industrial domains, conflict is not eradicated in an easy equilibrium between the two forces; indeed, conflict appears as a necessary part of social life because it protects people from the excesses threatened by the wholesale adoption of either social force (e.g., anarchy or totalitarianism).

Cooperation considered as a process of production, then, does not consist only of "interdependence" or "fusion." Instead, it consists of the interplay between integrating and disintegrating forces, as people drive themselves to forge connections even while preserving the very real differences that separate them. People struggle with these differences and the contradictions they reveal, but their efforts are focused less on resolving the contradictions than on containing them at every moment so that the social system neither flies apart in a centripetal frenzy nor comes to a halt in a centrifugal malaise. In this oscillating dynamic, conflict and harmony appear as contrasting values that can apply to either social force. As A. L. Epstein (quoted in Nader 1990: 301) notes, "living in amity is a social value, and different societies attach different weight to it." And not only different societies—what I suggest here is that amity, or "harmony," is a situationally invoked value. Cooperation considered analytically is not characterized by harmony alone, even if, ideologically speaking, people believe that it should be. Harmony is only one possible cultural trajectory within a more complex and dynamic process.

Cooperation, Power, and Hierarchy

People working together on European space science missions, whether moving toward integration or toward dispersal, whether negotiating in conflict or in harmony, are nonetheless all trying, generally speaking, to get things done. "Getting things done" raises the issue of power (Giddens 1984: 283). Power is, in a way, what cooperation is all about in an elementary sense, since as Giddens (ibid.: 31–32) argues, power is "inherent in social association (or . . . in human action as such)." Power is not, moreover, "inherently noxious," or even "inherently divisive" (ibid.: 283); it is simply, in this view, "the capacity to achieve outcomes" (ibid.: 257). Cooperation in these terms is a process that makes power available in collective action directed toward the production of artifacts, whether material or social; indeed, it is a particularly powerful form of human association.

One question that arises when such collective action is considered is how to realize it, pragmatically speaking; people wonder about how they can organize differences (of individuals, groups, expertise, components, organizations) to make production possible. Such a question raises issues of hierarchy, equality, and difference, issues that have preoccupied many anthropologists in different topical areas and different theoretical traditions. There is, indeed, a great deal of literature on this topic (e.g., Brenneis and Myers 1984, Britan and Cohen 1980, Dumont 1970 and 1986, Flanagan and Rayner 1988, Flanagan 1989, Greenhouse 1992, Howell 1984, Kapferer 1988, and Strathern 1987), but as in other areas of anthropological analysis there is little consensus on terminological definitions, let alone on theoretical understandings. I am not interested here in proposing a new theory of hierarchy, or of reviewing the rather unwieldy and disconnected literature on the subject. Rather, I want to define some of the terms I use to address the issues of hierarchy and power in the cooperative practices of ESA space science missions.

One of the difficulties in analyzing hierarchy in this ethnographic context is the rather strong views on the topic held by participants themselves. Over and over again, I heard participants reject hierarchy in no uncertain terms; they insisted, instead, on the egalitarian nature of their social process (see chapter 5 for a consideration of these views). Yet at the same time, hierarchy did play a role and was rather unproblematically accepted as the most logical means of organizing the differences inherent in the division of labor toward productive ends. I suggest that there are at least two understandings of hierarchy at work here, understandings that also mark anthropological writings on the topic.

On the one hand, there is hierarchy as a practical-technical principle of organization, one that reflects distinctions among tasks and the mutual dependence among them. The ordering of tasks (and groups and components) along a vertical axis is neutral, in the sense that there is no necessary valorization, culturally speaking, of higher and lower levels of organization. Indeed, the principle of hierarchy in this case is integration: lower-level tasks, components, individuals, and groups are integrated at increasingly higher levels of organization. The differences that define these elements are not eradicated in the process of integration; indeed, their differences make sense only within the context of the whole (see Piaget 1970 for a discussion of structure as a hierarchical organization of transformations). This is, moreover, not necessarily a hierarchy of "control"—which Giddens (1984: 283) defines as "the capability some actors, groups, or types of actors have of influencing the circumstances of actions of others." It is instead a hierarchy of power, in which the more integrated, higher-level elements have, by definition, a more extensive ability to get things done because they include in them the lower-level units without eradicating these units' capacity to get things done as well.

By describing the structural logic of this kind of hierarchy I do not mean to suggest that there are no values attendant on its implementation in processes of cooperation. This logic of organization is often ideologically supported by the belief that without hierarchy there would be no order, hence no production; there would be only limitless conflict, differences contending forever in unending chaos (Kapferer 1988). Particular values are, therefore, associated with this kind of hierarchical organization; in European contexts, the associated values are those of efficiency, pragmatism, utility, and harmony. Such an ideology not only supports production but can legitimate the status quo, that organization of differences which, those in the upper echelons can argue, helps to keep societal chaos at bay.

It is at this point that another idea of hierarchy insinuates itself. There is also hierarchy as a "system of values by which individuals and groups acquire or lose social and moral merit in the eyes of others" (Greenhouse 1992: 250). This is not a practical-technical hierarchy, but a political-cultural hierarchy, one that speaks directly to and about issues of control. In the eyes of ESA participants, political hierarchy is entirely unacceptable, because it results in domination, or the oppression of some people by others. This violates participants' sense of the inherent equality of persons, because political hierarchies not only rank but differentially valorize people. It is therefore directly about judgment and about control. It also, interestingly, violates participants' commitments to their differences; they see in the domination of political hierarchies an imposition of conformity as well.

Participants counterpose to this kind of political hierarchy a principle of egalitarianism, an egalitarianism they define as characteristic of their social organization and as a value in itself. As a social form, egalitarianism denotes a system in which there is "relative political autonomy of actors . . . ; [and] no defined relations of subordination or superordination" (Myers and Brenneis 1984: 11).[3] In the case at hand, there is no single encompassing, uniform, hierarchical structure of statuses or tasks that can be imposed on the multiple differences which characterize the social groups involved in cooperation. Indeed, cooperation understood as the negotiation of differences appears as just the kind of process necessary for production in such a social field *precisely* because there is no singular hierarchy by which tasks and groups can be easily subsumed and directed. This egalitarianism in the social field is matched by an ideology that declares diversity as its defining value; diversity is what characterizes this social system composed of different and equal actors.

Both hierarchy and equality are valued in working together (although participants find it hard to articulate their acceptance of "hierarchy"), precisely because each, in appropriate contexts of action and interpretation, offers a model for insuring the preservation of difference. This is indeed what people worry about—how to keep their differences, and hence themselves, alive (see chapter 6 for more detailed discussions of such existential worries). From the perspective of egalitarian values, participants fear what *they* call hierarchy because they worry that "the individual will be consumed, obscured, and will lose its identity in more inclusive orders and that those who command such orders will negate the autonomy of subordinates" (Kapferer 1988: 15). From the perspective of hierarchical values, however, the fear is different; here, the fear is of the individual "which asserts the determination of its individuality against encompassing and incorporating forces" (ibid.), forces that have established the interdependence of differences but not their eradication. The "aversion to hierarchy" (Dumont 1970: 239), then, is not a rejection of the efficiency or harmony promised in the logic of an integrating hierarchy, one that defines the order of their artifacts as well as the order of their social worlds (see chapter 7 for a discussion of this point); it is a rejection instead of categorization, of the erasure of differences that results from a political hierarchy of domination. Analytically speaking, participants are worrying that "hierarchy," which organizes differences in a vertical order of increasing integration, might turn into "stratification," which ranks human differences in institutionalized "categories of persons" (Flanagan 1989: 247–48; see also Kapferer 1988: 14 for this distinction).

The point I want to emphasize is that both fears are present in working together, and they are called up at different moments by particular contexts of action. There is, indeed, "a subtle and interesting interplay between hierarchy and equality" (Flanagan and Rayner 1988: 13) in this social system, as in others, which means, moreover that "we must withdraw from characterizing systems as either hierarchical or egalitarian" (Flanagan 1989: 262), and recognize instead the way in which relationships, statuses, tasks, and values can be organized according to both kinds of principles simultaneously and in a single social system. In the case of working together, moments of decision are particularly vulnerable to expressions of and reactions to such fears.

Decisions are significant in working together precisely because they are occasions that make manifest the problem of control in the exercise of power, as the technical hierarchy of integration threatens to collapse—in the sense that its structural complexity would be simplified—into a political hierarchy. In these moments, distinct participants arranged along a horizontal axis, their differences simply juxtaposed, find themselves divided vertically in terms not of task but of "control," as someone or some group asserts authority to which others must submit. These are difficult moments. On the one hand, as Kapferer (1988: 110) argues in the case of Sri Lankan hierarchy, this "is not mere submission, but a positive and forceful submission which demands the ordering reciprocity of hierarchy" (I would substitute, in this cultural context, "cooperativeness" for "reciprocity"). On the other hand, as Kapferer (ibid.: 171) argues in the case of the Australian ideology of egalitarianism, people resist such submission because egalitarian values are accompanied by "a form of self-effacing humility whereby individuals do not assert the dominance of self over other in accordance with their social identities in the nonnatural world." What in the case of a hierarchy of integration is simply the exercise of authority to get things done, in the case of political hierarchy appears as authoritarianism in order to get people to do things (see Greenwood 1988); this authoritarianism is unacceptable in either case. In order to get things done, then, people engage in the constant undermining of hierarchies, contesting assertions of control and constantly negotiating differences in an effort to release the emancipatory, rather than the constraining, power (Giddens 1984: 257) of cooperation.

Cooperation, Europe, and Diversity

As I have indicated above, participants working together on European space science missions treat "diversity" as a key value and principle of their social system. The principle of diversity mandates the assertion and protection of difference; however, it was not just any differences that were significant in the context of ESA missions. Participants recognized in particular two primary categories of difference—national affiliation and occupation (I discuss these differences ethnographically in chapers 3 and 6). These are, in a fundamental way, political categories, in that the significant social differences were defined by and emerged in the context of European political-economic cooperation.

That national affiliation should be a significant marker of difference makes sense in this process. The process of cooperation is, after all, undertaken in terms of the political-economic division of labor that characterizes European integration, and participants are drawn into working on space science missions precisely because of their citizenship in particular Western European nation-states. It is an open question, of course, to what extent such citizenship translates into meaningful identities, although participants tend to collapse these two dimensions in the ubiquitous term *nationality*; this is the sort of question that has interested anthropologists and other social scientists investigating national identity in Europe (e.g., Anderson 1991, Boerner 1986, Borneman 1992, Cohen 1986, Grillo 1980, Herzfeld 1987, Macdonald 1993, McDonald 1989, Tonkin, McDonald, and Chapman 1989, Sahlins 1989, Wilson and Smith 1993), and that I consider briefly here.

That *national* affiliation (as opposed to regional or ethnic identity) should be significant in ESA makes sense historically, ideologically, and structurally. Nation-states have been significant players in modern European history, and they have competed and fought with each other in the name of various nationalisms for a long time. National identity is continually brought to the fore as people struggle to protect, in different ways, the validity of nations, whether through the contemporary maneuverings in and resistances to the development of the European Community (the vote by the Danes against the Maastricht Treaty in 1992 being only one recent example), or more tragically through violence against people conceived as violating a sense of "nationness," as in recent attacks on Turkish workers in Germany or in the war between nations unleashed in Yugoslavia in the aftermath of communism's fall. Given this history of antagonism and confrontation, the fact that in ESA people from different European nations can work together, side by side, successfully and largely without recrimination is, for participants, constantly remarkable.

The pressure to notice and take account of national identity in ESA also makes sense in relation to what Herzfeld (1987: 78) calls the "European ideology" of unity in diversity, an ideology that has "proclaimed external distinctiveness as the historical destiny of the state, while reducing internal distinctiveness to the measure of that same common denominator." In discourses of national identity, differences among the citizenry—differences of region, ethnicity, race, religion, class, or gender—are defined as internal to the state, and they are, as Herzfeld argues, "anathema, inasmuch as they challenge . . . the authority of the state"; by contrast, "at the level of the state and above, cultural differences [i.e., national differences] . . . [are] 'European' and therefore good" (ibid.). This reduction of salient or meaningful differences seems to have been particularly effective in the case of intellectuals and bourgeois classes (the classes from which the professional participants on ESA science missions are drawn), those who readily identify themselves, at least when engaged in such transnational activities as academic research and international finance, in terms of their "nationality." This reflects the class interests of the nation-state, and of the "superstate" of Europe in the EC version (Galtung 1989).[4]

The effacement of discourses of regional or ethnic identity in ESA, then, reflects the terms of the European ideology, and this ideology is enshrined in ESA's organizational structures. It is a salient aspect of ESA in part because the Agency itself took shape in the context of the contemporary political movement toward European integration (see chapter 2), a movement seeking to knit some of the Western European nation-states together so as to make the enmity and prejudice of nationalisms untenable. Those working to unify Europe politically and economically in the form of various "supranational" European institutions, such as those of the European Community and ESA, hoped that these "would provide the working basis for . . . a gradual transfer of allegiance and loyalty . . . among the citizenry of Europe" from national interests to European ones (Bull 1993: 27). Such a hope indeed represented the "neo-functionalist dream" that "Europe will be . . . united by experts [and] by economic self-interest" (Townshend 1980: 197).

No doubt, experts *have* been united—this is, at least, a popular cynical view of European unity, as described for instance by Varenne (1993: 224–25):

> Quite common in conversations with my neighbors in Dublin . . . are statements to the effect that "Europe" is only a matter of interest for businessmen and bureaucrats, that nobody really cares whether something or somebody is European or not. In such conversations, the talk then proceeds to a reaffirmation of the reality of "Ireland," "Wales," "Scotland," "England," "France" and such more or less traditional symbols.

European leaders have, in response to such cynicism and resistance, turned their attention to the problem of how to generate an affective commitment to the idea of Europe, in effect acknowledging that political and economic efforts toward integration will not alone bring about allegiance to a new locus of power or generate feelings of belonging. To this end the European Community has created a flag, a symbol, and an anthem (see Townshend 1980), but these "top-down" attempts so far do not seem to have had any significant effect in generating widespread enthusiasm for a European identity.[5] Even for the "experts" at ESA, these objects do not generate a meaningful "European" identity; this is not to say, however, that these professional scientists and engineers do not articulate a European identity in certain contexts, as I have argued elsewhere (Zabusky 1993a and 1993b) and as I briefly discuss in chapters 3 and 6.

The elevation of a national identity over more regional or local identities is an effect of the specific context—historical, political, structural—in which these particular social actors find themselves. I am not, in other words, arguing that national identity is *more* important than other kinds of identities; as Wilson (1993: 17) notes, "it should not be surprising if most EC citizens recognize their own membership in a number of culturally defined groups—most people in the world do—and do not feel that these overlapping identities are incompatible." What is at issue are "the critical moments when one or more identities take precedence over the others" (ibid.). Discourses of identity—what is significant publicly, in a relation between people (Epstein 1978)—take shape in particular structural contexts and in turn inform the way in which these structures are themselves ordered. In this view, national identity is a practical identity, made real in this environment of European inter*national* cooperation, where it is nation-states that have made collective work on space missions possible in the first place.

National affiliation (a category that links legal citizenship and cultural identity) is not, however, the only important kind of difference acknowledged and contested by participants. Occupation is an equally significant category of difference in ESA space science missions; it is also, as in the case of national affiliation, a political category of difference (see Zabusky 1993a for a discussion of the way in which discourses of nationality and occupation are complementary). It represents, among other things, participants' practical awareness of the division of labor as an integral part of political-economic relations in contemporary industrial capitalism. As Durkheim (1933) noted, people in industrial societies come to derive their sense of individuality and uniqueness from their specialization in labor activities. Such specializations represent differences not only of technique and expertise, but also of class. Indeed, the awareness of class, along with the persistence of class-based values, beliefs, and practices, is an everyday part of European citizens' perceptions of society and social relations; it is in part through class that people recognize themselves and distinguish themselves from others (Bourdieu 1984).

The differences between the white-collar professions of scientist and engineer and jobs of secretaries and technicians are only one source of such class- and occupation-based diversity. Scientists and engineers, theoreticians and hardware specialists, pure and applied scientists are also divided from each other along lines marked by class. According to Berman (1981: 58), such divisions have a long history in Europe. Engineers, those who work with mechanical things, and those who produce "useful" objects are not, in this cultural taxonomy, engaged in the "clean" work that characterizes intellectual pursuits. Given the historically "strong intellectual bias against craft activity, with its lower-class associations" (ibid.), a bias reaching back into medieval times, people who engage in such activities therefore bear the inescapable stain of lower-class mores.

Shapin (1989), examining the florescence of natural science in seventeenth-century England, concurs with this view, noting in particular that the "moral economy" of the times drew a distinction between "mere skill (or hand knowledge)" and "truly reflective and rational philosophical knowledgeability" (ibid.: 561). This distinction reveals "the traditional contempt that genteel and polite society maintained for manual labor" (ibid.), a contempt that rendered laboratory technicians "invisible," in Shapin's view, and continues to affect contemporary relationships in laboratories and organizations where technical and scientific work is the main activity. Thus the meaningfulness of occupational diversity, which is such a significant factor in the social system of working together on space science missions, contributes to a political perception of important social relations, for it relates to the class distinctions that accompany the division of labor in capitalism (see also Bourdieu 1988 for a discussion of class, politics, and academic professionals).

As I have argued here, the diversity that is so important to European participants on space science missions is a diversity of social categories that are understood to be, in a fundamental way, political. As such, the meaning of diversity does not extend to social categories that are understood, at least in European and American cultural perception, as being based in biological difference. For this reason, categories of, for instance, race (a matter of "blood" and of "bodies"; see for example Dumont 1970, Greenwood 1984, Räthzel 1991, and Segal 1991) and gender (understood as deriving primarily from biological sexual differentiation rather than social differentiation; see Ortner 1974 and Ortner and Whitehead 1981) are excluded from consideration. As a result, the absence of racial or gender diversity among people involved in ESA space science missions was largely unnoticed by participants.

For instance, at the time of my study (1988–1989) at ESTEC, there was only one prominent, senior engineer who was of African descent (originally a citizen of Jamaica), and only a small number of engineers of Asian descent. All the scientists were white. Nonetheless, the absence of blacks, East Asians, and South Asians was not meaningful for participants; that nearly everyone was white did not seem to contradict people's primary experience of diversity or to challenge their corresponding valorization of diversity. The silence about race at ESTEC finds its match in a silence about race in what McDonogh (1993: 144) calls the "sophisticated pluralist models" of confederation that characterize the political efforts at European integration, particularly in the EC. McDonogh notes that this movement raises practical and cultural questions about the boundaries of citizenship—about inclusion and exclusion—that are being formulated in a political contest which takes into account national and European ethnic groups but not, or not always, "the relationships of European states with colonials (and former colonials) and the presence of significant populations of color within Europe" (ibid.: 145)—except perhaps to devise various ways, physically or legally, to exclude members of such populations (e.g., Lloyd and Waters 1991, Pieterse 1991, Webber 1991). In this way, the idea of race silently, even invisibly, contributes to the production of a new "European" nationality in the political-economic space of European integration (Segal 1991).

There is a silence, too, about gender in the context of ESA space science missions. At ESTEC, at the time of my study, there were no women engineers, no women technicians, and only a small number of women scientists (all in postdoctoral positions) and computer specialists, although a few women engineers or scientists, affiliated with external institutes, would participate in missions from time to time. Nonetheless, the fact that nearly every professional was male did not seem to undermine the experience of and value placed on diversity for and by participants.

This invisibility of gender as a significant category of difference characterizes also the discourse and structures of other European institutions. Provine (1993), for instance, offers an interesting critique of the European Convention on Human Rights for just this omission. She notes that the European Commission and Court of Human Rights, the adjudicating bodies established to hear and resolve complaints arising under the convention, do not compile statistics on the sex or race of applicants; "only nationality, and upon occasion, citizenship are noted in the yearly profiles of filings and decisions" (ibid.: 20). These criteria reflect, in an altogether different context, the emphasis on what those involved consider to be significant political categories of difference, and their concomitant discounting of biological ones. Provine indicates that women are absent not only statistically, but culturally, in the very construction of what rights are legitimate and what rights can be admitted as "human" by the convention. She notes that, "preoccupied as they must have been with the devastating experience of Nazi expansionism, it is understandable that human-rights advocates of the post-war era focused on controlling the capacity of the state to crush individuality and political freedom" (ibid.: 22–23). However, by conceiving of rights in this way, the framers ignored the fact that for women, "the common heritage [in Europe] is state-mandated exclusion from the political process. . . . Silence on this legacy of legal oppression suggests the insignificance of these realities to the drafters, and thus the insignificance of women's experience to the effort to establish a regime of human rights in Europe" (ibid.: 23). Women's experience has been, in this way, de-politicized; as such, it is not available as a salient category of difference in contexts that define difference in terms of political significance.

The omission of women (both physically and culturally) from participation in European space science missions must be seen not only in relation to the political economy of European integration, but also in relation to the social and cultural systems of science and technology. There is a vast and rich literature on women and science, a literature that takes up not only the statistical absence of women in science but the way in which science must be understood, historically and culturally, as itself a gendered institution (e.g., Haraway 1990, Harding 1986 and 1991, Harding and O'Barr 1987, Keller 1985, Martin 1987, Merchant 1980, Noble 1992, Rapp 1991, Rossiter 1982, Schiebinger 1989, and Traweek 1988). Given the analyses made by such scholars as these, it may come as no surprise that there are no women working on ESA space science missions, even though participants value "diversity"; the world of science was established specifically to be, as Noble (1992) argues, a "world without women." The gendering of science, according to Noble (ibid.: xv), arose in medieval times: "It is here, in the struggle of the Latin clergy to impose itself between God and the rest of humanity, that the curious culture which spawned Western science took shape," a culture that "has not simply excluded women, [but which] . . . has been defined in defiance of women and in their absence" (ibid.: xiv). This historical underpinning of science is replicated and reproduced in the doing of science; as Keller (1985: 8) argues, "ideologies of gender and science inform each other in their mutual construction."[6]

I do not, in this book, pursue an extensive critique of this silence about race and gender in the context of European space science and technology, which is not to say that such a critique is unnecessary. Indeed, in my view, it is deeply necessary; nonetheless, my goal here is simply to point out the bounds of diversity in this ethnographic context. My purpose is not to argue that the absence of these differences reveals that in ESA there is "really" no diversity at all, or that diversity is not "really" a value, but to make clear the meaning, the explicitly political meaning, that the concept of diversity has for participants. As I have suggested, when we take into account the historical and cultural background within which European cooperation in space science is undertaken, it is no surprise to find particular categories of difference coming to the fore. Nonetheless, I hope to suggest the limitations of this concept of diversity in allowing for certain kinds of change; as long as race and gender are not recognized as categories of difference with political import and meaning, they can never be included in a universe of discourse that by definition discounts their significance.

Cooperation, Science, and Technology

My focus on the development of space science missions in ESA connects this ethnography of cooperation to the well-established literature in the sociology of science and technology. Although this is an ethnography, it is not, as is so much of the recent literature on the sociology of science, a "laboratory ethnography" (e.g., Latour and Woolgar 1979, Knorr-Cetina 1981, Lynch 1985). This is so for several reasons.

First, most of these contemporary laboratory ethnographies of scientific activity are concerned primarily with epistemological issues relating to what sociologists of knowledge take to be the focus of scientists' work: the production of scientific facts and ideas. The primary motivating question of such studies is: How do scientists know what they know? Much of this work is geared to demonstrating how knowledge, including scientists' knowledge, is socially constructed (i.e., rather than given in nature). Fuchs (1992: 42–43) argues that such microstudies of scientific research activity "define themselves not so much in opposition to classical sociology of knowledge or orthodox Mertonian sociology of science, but in opposition to traditional realist philosophy"; in other words, the implementation of "ethnography" as a research methodology was geared to discovering what *actually* happened inside the "Pandora's box" of scientific research (ibid.: 3), in an effort to challenge the idealized and empiricist depictions of scientific method essayed by positivist philosophers (such as Popper 1961; Fuchs 1992 and Hagendijk 1990 offer particularly cogent critiques of the social constructivist approach to science from more sociological and structural perspectives).

My own study is not driven by epistemological concerns; science is, in this study, not a "mode of thought" (Horton and Finnegan 1973), but a social institution and a cultural discourse implicated in, influenced by, and influencing in turn political, economic, and organizational structures in which particular social actors who are called scientists try to carry out their work. My attention is turned, thus, to social processes that are about work and how to work, and not about scientific knowledge. In this sense, like Fuchs (1992: 7), I regard science as a kind of "work organization," rather than exclusively as epistemological inquiry. This shift in emphasis requires an approach that includes "putting science back in society," a task undertaken in diverse ways by the contributors to the volume edited by Cozzens and Gieryn (1990: 1); as should be clear from the discussions of theoretical and topical ideas in this chapter, I draw on the "literature of generic sociology [and anthropology]" (ibid.: 12) in my investigation of the practice of cooperation in the production of space science missions for ESA.

This emphasis signals a second way in which my study departs from sociological accounts of "technoscience" (Latour 1987).[7] Like other anthropologists who have recently begun studying contemporary science and technology in earnest (Hess and Layne 1992), I pay attention to the "broader cultural arenas" (ibid.: xii) in which scientific and technological work is carried out. This is because my own approach is founded in "the cultural perspective, critical questions, and ethnographic methods as they have been formulated in contemporary American (and to some extent Anglophone) social/cultural anthropology" (Hess 1992: 2). Ethnography from this perspective necessitates "talking to and interacting with people, and ultimately attempting to understand their symbolic worlds and *social action*" (ibid.: 4; emphasis in original).

The cultural approach that I take, moreover, does not mean attending only to "the technical content of scientific knowledge" (Cozzens and Gieryn 1990: 7), which is how sociologists of science often define *culture*; it means trying to understand what the world, including the social, political, and organizational world, looks like to participants. This effort requires attending to, in Hess's (1992: 3) playful revision of Malinowski's dictum, "the natives' *points* of view," in full recognition of the fact that culture is "multivocal, fragmented, and contested" and, I would add, dynamic. This approach to the ethnography of technoscience demands that analysts consider the wider political, organizational, social, and cultural contexts within which, and in terms of which, this work of technoscience is carried out. In this way, along with other anthropologists, I am "decentering the laboratory and transforming it into just one of the 'scenes' or 'sites' for the ethnography of science and technology" (ibid.: 14). I do not, and cannot, separate scientists from the heterogeneous environment in which they work; the process on which I focus in this book concerns not scientific research contained in a more or less bounded environment (such as a laboratory), but instead the development of scientific missions, which includes scientists, engineers, administrators, technological artifacts, scientific ideas, and political and economic resources.

Finally, there is a third way in which my work differs from the prevailing approach in the social studies of technoscience, and that is my emphasis on "cooperation." I do not choose this emphasis in order to resurrect the idea of an altruistic and consensual "scientific community" as a sociological object. This was the notion informing many classical studies of science (e.g., Hagstrom 1965, Crane 1972, Merton 1973, Mitroff 1974, and Price [1963] 1986), and it has been explicitly rejected by sociologists since the mid-1970s (e.g., Bourdieu 1975, Collins and Restivo 1983a and 1983b, Fuchs 1992, Latour 1987, Latour and Woolgar 1979, Mulkay 1975). These days, sociologists privilege "conflict"—including competition, disagreement, self-interest, and aggression—in their studies of science and technology. Above, I considered a similar emphasis in anthropology; in this section, I note the specific theoretical context of this view in the sociology of science and its implications for analysis.

The recent preference for attention to conflict over the Mertonian focus on community and consensus takes many forms. Mulkay (1975), for instance, takes to task the very idea of the scientific ethos described by Merton, arguing that scientists' articulation of such norms should not be embraced as an accurate description of social rules that "control" scientists and thereby produce "conformity" among them. Rather, these norms are "vocabularies of justification, which are used to evaluate, justify and describe the professional actions of scientists" (ibid.: 654). In short, Mulkay identifies not a scientific community but "an interest group with a dominating elite, and a justificatory ideology" (ibid.). Bourdieu (1975: 19), writing from a different sociological tradition, also rejects the "irenic image of the 'scientific community,' as described by scientific hagiography—and often, subsequently, by the sociology of science"; instead, he argues that "the scientific field is the locus of a competitive struggle" (ibid.), in particular "a political struggle for scientific domination" (ibid.: 22). Similarly, Collins and Restivo (1983a: 196) have argued for an understanding of science that "stresses political processes operating in 'career spaces.'" In their study of mathematical research from the early 1500s until today (Collins and Restivo 1983b: 222), they depict mathematics as an arena in which "ambitious intellectuals pursuing self-interested paths to fame and fortune took advantage of whatever organizational resources new situations offered." In this view, intellectual advance depends on controversy and competition, not consensus and cooperation, as in the Mertonian vision. Finally, many social constructivist studies of scientific research, having discovered that "negotiation" is critical to the production of scientific facts (Fuchs 1992), tend to stress the aggressive and oppositional elements of such a social process. For instance, Latour and Woolgar (1979: 243) portray scientific research activity as "a fierce fight to construct reality," in which competition and agonistic struggles are key in the production of facts. Latour (1987: 172) goes so far as to depict technoscience as "part of a war machine" that produces scientific facts through "the mobilisation of resources, whereby I mean the ability to make a configuration of a maximal number of allies act as a single whole in one place."

This emphasis in the social studies of science has made it difficult to talk about community at all; to do so leaves one open to charges of being naive at best or, at worst, a functionalist. Nonetheless, as Fuchs (1992: 8; emphasis in original) observes, "this critique has gone too far in abandoning *any* notion of community organization in science in favor of a radically situationist and actor-centered microposition." Hagendijk (1990: 50), similarly, deplores social constructivism's "overly voluntaristic view of science and . . . rejection of explanations of scientific change that invoke the existence of a social and cognitive order going beyond what specific participants in a given interaction may specifically mention."[8]

In this ethnography I am, in fact, more interested in the scientific community than in science per se, in the minimal sense that I am turning my focus to social relationships rather than to knowledge. This does not mean that, like Merton, I am simply trying to describe the norms and values of an extant community defined unproblematically (see Traweek 1992: 437–38 for an ironic comment on this view of "community" and "culture"). What it does mean is that I am trying to understand, among other things, how such a community constructs itself, and how the idea of such a community affects the process of working together (see, for instance, the discussion in chapter 7). This process indeed depends on the construction of a social system that incorporates or cuts across an entity which participants call "the scientific community."

My looking at community, then, does not result in my seeing only conformity; what I am trying to do is find a more positive way to talk about the "negotiation" that is an essential part of any human social endeavor, whether science or politics or anything else. As I see it, there are no necessary connections, in sociological terms, among the ideas of community, cooperation, harmony, and conformity, although the sociology of science tends to make these equations. "Communities," as anthropologists well know, are full of conflict; this may not make them any less "communities" to those who live in terms of them. Furthermore, for people to conceive themselves as being part of a community does not suddenly make them automatons, obediently following social rules, whether of scientific method or of dietary etiquette (Comaroff and Roberts 1981 address just this issue in their discussion of rules, processes, and conflict).

I talk about "cooperation" in order to counteract what I see as the prevailing tendency among sociologists of science interested in "conflict" to depict science as a particularly aggressive and oppositional form of social interaction (as "war continued by other means") at the expense of any more positively construed associative impulses at all.[9] I do not dispense with the essential insights gained from the social constructivist approach to science (e.g., that scientific knowledge is socially produced); rather, I propose to integrate them with a more structural and cultural view of social action, with a view which acknowledges that people do share things. Nonetheless, my focus on cooperation does not take agreement for granted (nor, for that matter, does my focus on cooperation take conflict for granted); it seeks to understand how it is that scientists and engineers involved in space science missions come to agreement, and, critically, how they take such great pleasure in this process, even as they are riven by all manner of differences. Solidarity can be understood, in other words, in a positive sense, not simply as a defensive tactic or a strategy of domination, images that depend in the first place on an understanding of human social life as militaristic and aggressive. Solidarity emerges, in fact, from the very act not only of talking together but of arguing together about things that matter.[10]

The Ethnography of Cooperation: Methodology Considered

The practice of cooperation that I explore in this book poses certain challenges to the anthropological method of participant-observation, the mainstay of ethnographic inquiry. One problem is that local practices sometimes seem entirely dependent on macro-level political and economic processes and structures (such as states and bureaucracies). Furthermore, the practices that make up "cooperation" do not happen in only one place, but in many places, simultaneously. What is a lone ethnographer to do?

I chose to try to understand the somewhat chaotic process of working together from a single vantage point, that of participating scientists, particularly those working in and for ESA. Practical considerations contributed to this decision—my limited funds and time prohibited extensive travel; moreover, constant travel to multiple sites would have undermined my ability to get to know anyone well enough to find out what they thought was "really" going on, as opposed to gathering various documented "official" versions. But situating myself in this way also made ethnographic and theoretical sense. For one thing, by choosing a single site from which to observe and learn about cooperation, I acted like any participant in working together; people can only participate in, and come to know well, one node of the complex, interlocking process that is working together. In any case, from the participants' perspective, there is no place from which to get a definitive reading of what is going on, since the social field of working together depends on egalitarian diversity and the interdependence of distinct participants. Although all are involved in the same process, given participants' multiple affiliations everyone has a different perspective on the work at hand. Choosing a place to be, and to belong, is indeed part of working together.

This makes sense theoretically as well; there cannot, in fact, be a "comprehensive" perspective on working together from anywhere inside the process. To get an overview of working together would require situating oneself in the domain of structure, if such a thing were possible (perhaps if one were to take a ride on an orbiting satellite . . .). But structure, as I have argued here, is not something one can inhabit; it exists only as it is instantiated in the experiential, practical, and local here and now, and it is precisely participants' perspectives on structure that I endeavor to convey here by interpreting their practices. This practice is itself inevitably partial because it happens through time and in specific spaces; it consists of interruptions, ambiguities, and negotiations of meaning. To be inside practice, in other words, is to be situated somewhere and somewhen in particular; but to be inside practice is also the only way to come to know, ethnographically, the social system, since the system is replicated and reproduced through agents' everyday actions, evasions, and enunciations of value.

As I indicated above, I chose to focus on the scientists; these participants are themselves a dispersed and diverse group, and I chose to focus on the experience of a particular subset, the staff scientists of ESA. Accordingly, I settled myself (from September 1988 to August 1989) into the Space Science Department (SSD) at ESTEC, the largest ESA site. I want to underscore here that, although I was based in SSD, this book does not present a study of the "organizational culture" of SSD, of ESTEC, or of ESA. I studied, to paraphrase Geertz (1973), not *an* organization, but *in* an organization. Indeed, I chose SSD because it seemed to serve as a kind of "crossroads" during mission development, one that opened out to and admitted other locations and social groups. In this way, it provided a complex vantage point from which to observe working together, because the staff scientists played a critical "liaison" role in the social system. They functioned as the "interface" between academic scientists (who are the "users" of ESA space science missions) and the mission engineers in ESA. Because they were thus engaged with both scientific details and engineering concerns, they were able to keep in contact with those doing work on scientific instruments in universities and with those doing work on spacecraft technology in industrial firms, as well as with the ESA scientists, engineers, and administrators who helped to coordinate mission work.

My choice of ethnographic location necessarily influenced my observations and understandings of the practices of working together. For instance, I frame my questions and my answers in terms of the scientists and their view on the complex processes in which they are involved. Had I situated myself inside an engineering team, for instance, I suspect that talk about science and interactions with scientists would play a relatively minor role (even though these are scientific missions), while the talk about economic concerns and interactions with industrial engineers would loom correspondingly larger. Furthermore, because the activities of ESA staff scientists were focused on liaison work, they shared, and I with them, an overriding concern with the problems and possibilities inherent in making and finding connections.

Although much of what I describe in the book, then, reflects the particulars of the view from SSD, my ethnographic inquiries were not confined to the operations of that Department, nor to the scientists who make up the majority of its members. I could not have such an exclusionary focus, given the dynamics of working together. Of necessity, I spoke at length with scientists and engineers in universities, ESTEC engineers working on mission teams, and other SSD staff (including engineers and technicians); these are the same people with whom SSD staff scientists speak in their mediating role in working together (I describe these various participants in more detail in chapter 3). I met these people at ESTEC, when they came for meetings and design reviews, but I also traveled to external sites on several occasions to conduct interviews and attend meetings. It was by listening to as many different perspectives as possible that I learned about the complexity of working together—even while I spent most of my time in the halls and offices of SSD—since as I have suggested scientific missions look quite different from other venues, whether that of rank-and-file engineers on mission teams or that of academic scientists developing instruments in university laboratories.

My field research was facilitated by participants' common use of English both in work-related situations, such as meetings, and in informal interactions; occasionally, my knowledge of French and German did prove useful, especially in informal settings. English and French are the official languages of ESA, and although all staff members are expected to know both languages, for most scientists and engineers English is preferred. Scientists in national universities are also conversant with English, since this language has been the lingua franca of international science since the end of World War II; many are able to speak quite fluently and easily in English, even though this is not the language of daily work in their home institutions. (For some discussion of the role of multiple languages in working together, see chapters 3 and 4.)

Another type of language problem was presented by the technical language of space science research and technology. Although I had embarked on my field research equipped with a basic course on astronomy, I found that this prepared me only for discussions on certain research topics, discussions that were relatively infrequent. My knowledge of astronomy was most useful, for instance, during the biweekly scientific seminars in SSD when the speaker was an astronomer and not a researcher in the other disciplines of "space science." In-depth discussions about the technology, however, constituted the majority of conversations and meetings, and these were often difficult for me to follow. As I discovered, however, because of the participants' diversity I was often not the only auditor who was left in the dark by technical discussions. Participants' specific foci left them ill-equipped to understand the intricacies of others' instruments, software, or even research ideas. Indeed, it is just this lack of understanding, an effect of specialization in the division of labor, that cooperation in work is intended to overcome. Over time, and with generous help from many participants, I came to understand the outline of technical problems and issues, and so could follow presentations, debates, and arguments in their general substance if not in their details.[11]

I collected data by a number of methods. I conducted an extensive series of semistructured and unstructured interviews with a wide variety of participants. In these interviews, participants and I talked about technical details—for instance, people might explain to me the subject of a plasma physics experiment, or they might discuss the physical and technological elements of the design of an instrument—as well as the social mores and politics of working together. I combined such formal conversations with more informal ones that took place at social events of many kinds. There were special occasions such as birthday parties, celebratory parties, and farewell parties held during office hours at ESTEC or in the evening at local restaurants. And there were also the more mundane, everyday interactions by the coffee machine or in the cafeteria over lunch. In these moments, I was not a passive listener but an active participant.

Published material and other written documents provided a wealth of data that complemented the information I gleaned from formal interviews. I read constantly and extensively; I perused technical documentation, meeting minutes, scientific papers, ESA publications, and general press material on space. These materials offered detailed information about missions, instruments, and the organization of ESA; they also expressed the rhetoric of cooperation so often heard in the milieu of European science. And because these materials usually concerned issues of current interest to participants in working together, they often provided conversation points for me and my informants, thus helping me to understand the perspective of participants on events and policies taking shape outside the immediate environment.

In addition to interviewing, participating, and reading, I also observed. On occasion, I had the opportunity to be an observer of engineering sessions in the laboratories of ESTEC, when engineers and scientists worked together to integrate and/or test equipment and software. For the most part, however, my observations focused on meetings; those I attended included both departmental staff meetings and mission-oriented team and working group meetings. Mission meetings in particular play a key role in working together, since attendees include not only SSD staff, but engineers and scientists from diverse locales. Some of these meetings lasted an hour; others stretched on for days and combined formal discussion with socializing over meals, when I would be transformed from observer into participant. At times, I even became the subject of interrogation in turn, as the scientists and engineers whom I interviewed and observed at work grilled me about my intentions, methods, and hypotheses.

Most people accepted me as simply another "researcher," another scientist, conducting work of a kind similar to their own, if in a different domain. Nonetheless, some SSD staff members remained uncomfortable with my presence at meetings and mealtimes, fearing that I was there at the behest of "management" to conduct some kind of "time-motion study." As a result, some were reluctant to talk to me about their daily activities, or to include me in informal conversations. On occasion, people asked me, jokingly yet nervously, whether I was "keeping track" of who was attending meetings, of how long people stayed at lunch, and so on. Participants who were not part of SSD had other concerns. Their overriding worry was about "confidentiality": whether the comments and arguments I heard would be attributed to people, and whether participants would be recognizable. One external scientist worried that I would write an "exposé," embarrassing those involved; this had apparently happened to him and some colleagues at the hands of an inquiring journalist. In response to such fears, I repeatedly made assurances that I would not share my notebooks with anyone, and that in my writings I would not use any names and would make every effort to disguise specific individuals. In this book, indeed, I use pseudonyms, downplay national affiliation as a descriptive feature, and occasionally disguise the gender of speakers (because there are so few women) in order to protect participants' confidentiality.

For the most part, however, people were deeply interested in what I would discover about them. Participants' interest in my study manifested itself particularly in repeated inquiries about what I was "finding out." On several occasions, I was asked during a meeting coffee break whether I would be presenting the results of my observations of that meeting at the end of the day, and many people asked me to send them copies of papers or of the book once it was finished.[12] Their interests indeed matched my own, and as a result my conversations with participants often took the form of mutual exploration rather than unidirectional examination, as we together puzzled through the intricacies of the practice of cooperation.

Two

The European Space Agency and the Structure of Cooperation

> The Agency itself, with its staff and committees made up of representatives of the Member States, constitutes one of the melting pots for the material from which Europe is gradually being forged, and in which nationalist preoccupations have to give way to wider, more promising visions. All who contribute to the life of the Agency have a sense of belonging to a European unity, the existence of which is indispensable to the fulfillment of those ambitions which the countries of Europe may have in the space sector, and cannot fulfill on a national level.
>
> *(From Twenty Years of European Cooperation in Space* [Longdon and Guyenne 1984: 229])

ESA IS ONE of the most significant instantiations of the structure of cooperation to which people have recourse in their daily practices. Through the various political agreements and bureaucratic regulations that brought it into being it, ESA establishes the ground state of working together by serving as its instigator. It provides the social, cultural, and sometimes physical space within which members of a variety of groups can come together to work on space missions. In this way, ESA creates both the context of and the medium for local negotiations, simultaneously delimiting the scope for action and providing many of the significant terms through which people render their practices mutually intelligible.

While ESA thus offers a forum for interaction, it cannot, and does not, dictate action; the differences that it brings together unfold with their own dynamic in the practices of participants. Indeed, from inside practice, the cleanly identified structures of ESA—structures that incorporate the history, politics, and economics of its establishment—appear to dissolve; people manipulate, challenge, and otherwise call into question the solidity of cooperation as they struggle to find ways to work together. These struggles themselves make ESA real; that is, ESA can be understood not only as a context for action, but as the outcome of a set of practices of negotiation, in particular of the negotiation of the differences that constitute the material out of which cooperation is forged.

Despite the dialectical relationship between the structures of cooperation and participants' local practices, the scientists and engineers who were engaged in local, technical activities were inclined to ask, "What does all this have to do with me?" when they heard about the politics of European integration, the economics of funding ESA, and the administrative tasks of coordinating a variety of participants. "I'm just doing my work," they insisted, denying the relevance of such structural parameters to their own practices. At the same time, however, they also recognized that the political-economic pressures on ESA in an integrating Europe, and the bureaucratic operations of an intergovernmental agency, did affect their efforts, even as they focused on the nuts and bolts of spacecraft design, the radiation tolerance of instrument components, or data tracing out colorful patterns on computer screens. "I'm just *trying* to do my work," they might say, qualifying their insistence that they were separated from the cooperative framework circumscribing and in some ways defining their efforts. In fact, although participants often expressed a sense of confinement and estrangement, they also acknowledged that without the cooperative framework of ESA, there would be no opportunity to work at all. In this way, ESA offers participants resources for sense making (culturally speaking) as they struggle to find their way over an ever-shifting ground of limits and possibilities in the realization of space science missions.

In this chapter, I examine the structural resources of working together as I explore ESA's history, rhetoric, and organization. In other words, I focus here on that framework within and against which participants are working. I begin with a brief history of the Agency, locating its genesis and continued operation in the political-economic movement toward European integration. Second, I examine the public discourse by and about ESA as an agency dedicated to cooperation, revealing a rhetoric of pragmatism that resonates with the ideology of European integration more generally. Third, I describe the organization that is ESA, reviewing its bureaucratic, economic, and political makeup. In the chapter's final section, I link the European and pragmatic cooperation that ESA instantiates to the problems and issues of cooperation in space science more specifically by discussing the Science Programme of ESA and the scientific missions that are its substance. I pay particular attention to these missions—what they are and how they are chosen, designed, and produced—in order to sketch out the details of the division of labor that underlies their construction.

Cooperation as Politics: The History of ESA

> The purpose of the Agency shall be to provide for and to promote, for exclusively peaceful purposes, cooperation among European States in space research and technology and their space applications, with a view to their being used for scientific purposes and for operational space applications systems.

These words from the Convention of the European Space Agency signed in 1975, which adorn the inside front cover of every *ESA Bulletin* (the quarterly publication of the Agency), reveal that the work of ESA is not the execution of research and development in space science and technology; rather, the work of ESA is cooperation. The emphasis on "cooperation" in ESA's mandate reflects the fact that ESA began its existence at a time when the movement toward European integration was gaining momentum.

The trend toward European union, resurgent in the aftermath of World War II, generated the formation of diverse organizations and groups, each committed, in different ways and in different areas, to fostering supranational integration in the interests of peace and prosperity. The EC is the dominant organization promoting this integration, but it is by no means the only one. In addition to this and other organizations with explicitly political and economic goals, a variety of "functional institutions with specific technical tasks" were established as well (K. Twitchett 1980: 19). These included such institutions as ESA and CERN, the renowned high energy physics research laboratory. Although scholar Kevin Twitchett (ibid.) argues that "political considerations as such are not normally within . . . [the] purview" of these institutions, proponents disagree, noting that these organizations play a critical role in the effort to forge political and economic unity in Europe, even though those working for them execute primarily "technical tasks." Chancellor Kohl of West Germany expressed this point of view clearly on the occasion of ESA's twenty-fifth anniversary, when he was an invited speaker at the gala celebration: "The joint European conquest and utilisation of space also strengthens the European identity, and this makes ESA's activities a major factor in building Europe as a political entity. . . . European unity is more than a matter of declarations. It must also prove itself in a field like space, which is so important for the future" (European Space Agency 1989: 20). ESA is not simply a technical agency, then, dedicated to science and high technology to the exclusion of all else; it is an active participant in the political and cultural arena where European unity is being constructed.

The Agency is self-conscious about its role on this European stage, and its own public discourse stresses the way in which the quotidian, technical work carried out by those working on space missions helps to realize the varied demands of political-economic cooperation. In the Agency's retelling of its own history in a publication issued on the occasion of its twentieth anniversary, the story goes like this:

> When they set up the European space organisations, the governments of the Member States were not just expressing their wish to be involved in avant-garde technology. They were also acting politically.
>
> They decided that, in a field full of promise for the future, European solidarity should take precedence over national effort, because that solidarity was a necessary condition for the effective participation of European countries in the exploitation of space technology. . . .
>
> The European Space Agency, then, represents a tool for working towards a united Europe—as it were, a 'European Space Community'. (Longdon and Guyenne 1984: 229)

The authors of this celebratory history of ESA here explicitly liken the work of ESA to that of the European Economic Community, substituting "space" for "economic." They thereby emphasize the way in which work in space science and technology is commensurate with work in political and economic fields when it comes to promoting European integration. Indeed, ESA appears here as cooperation incarnate; this is the message of its mandate.

ESA dates its beginnings from 1964, when two independent space-related organizations were founded following the onset of the space race between the superpowers. Historian Walter Laqueur (1982: 297) notes that both scientists and governments realized that "a divided Europe could not support the massive and specialized activities that were becoming an essential precondition of progress in certain fields." ESRO, the European Space Research Organisation, was dedicated to scientific missions, such as space probes, sounding rockets, and orbiting telescopes, designed to study properties of the local, solar-terrestrial environment as well as those of more distant planets, stars, and galaxies; scientists were active in bringing about the establishment of this organization, persuading various European governments to get involved (Krige 1992 and 1993). ELDO, the European Launcher Development Organisation, was dedicated to the development of heavy launchers, those rockets designed to lift large and heavy satellites into Earth orbit; national governments instigated this organization's founding.

As time went on, the European governments that had signed the conventions establishing both organizations became increasingly interested in pursuing more space missions that they characterized as "useful" (such as weather and communications satellites), and expressed their reluctance to remain committed solely to missions dedicated to scientific research. This desire, coupled with a faltering ELDO and an ESRO whose missions, though successful, were becoming increasingly expensive, led to a "crisis" in the life of regional space activities (Russo 1993a). This crisis did not lead to dissolution of the European endeavor into independent, sovereign space efforts, however; it was instead the catalyst for metamorphosis. In 1975, a new convention merged ESRO and ELDO into the European Space Agency. ESA emerged strong from this crisis of confidence, and the cooperation of European nations in space forged ahead.[1]

Cooperation as Rhetoric: The Role of ESA

ESA has indeed been undeniably successful in its space efforts—year after year it has launched (often on the European-made Ariane rocket) satellite after satellite, and the participating nations have reaped the corresponding economic, technological, and political rewards. However, as I have argued, ESA's role lies not only in the cultivation of high-quality technology, and in the development of successful space hardware and space missions, but in its symbolic participation in European integration. For both participants and observers of the European scene, ESA offers tangible proof that European cooperation can work. Indeed, the public discourse in and about ESA makes much of this "cooperation." In this section, I discuss just what cooperation means for those who point to its importance and its necessity.[2]

For many of those who talk about ESA and its link to, or manifestation of, "cooperation," the use of this term signifies not moral cooperation for peace but instrumental cooperation for profit. Cooperation is, in this view, "an efficient and cost-effective way for European States to contribute jointly to mankind's space adventure, thus participating together in the overall space venture" (Longdon and Guyenne 1984: 250). This kind of view is also evident in discourse surrounding integration in the European Community (EC), a discourse that articulates not only a "pacifist motive" (Holmes forthcoming) but a functional one behind the reasons for and consequences of European cooperation. These moral and functional rhetorics mingle as, for instance, in a quotation I repeat here from the Introduction, the assertion that cooperation among European nation-states will insure "a better, freer, juster world in which social and international relations would be conducted in a more orderly way" (Borchardt 1987: 6). The suggestion that cooperation might offer a "more orderly way" indicates that it is not viewed as just a moral symbol of peace but also as the rational thing to do.[3]

Although ESA may be representative of successful cooperation, then, participants and supporters do not regard "cooperation" as being particularly meaningful; it is to them simply functional—it is a practical method for achieving distinct ends. These ends are in themselves diverse, combining the concrete with the more intangible: artifacts on the cutting edge of new technology, economic wealth, political influence, regional integration. Cooperation in the ESA context does not offer a vision of possibility, nor does it represent a moral ideal; it is simply something nation-states do because it is useful, and profitable, to do so.

From an anthropological perspective, however, the focus by participants on "cooperation" as a functional mechanism is itself meaningful. That is, cooperation is expressive of a key cultural value for those who work for and with ESA (as it is for and with Europe more generally)—the value of pragmatism. Cooperation is in fact the embodiment of a pragmatic approach to the world. Such an approach emphasizes what is "realistic" and "down to earth," and opposes itself to what might be regarded as "mere" flights of fancy. A pragmatic attitude permeates ESA rhetoric. Absent are more romantic images, which Americans, for instance, might associate with space, not least because such images often color NASA publications and presentations: astronauts leaping on the moon, spiral galaxies spinning their star-filled arms light-years away, spacecraft speeding toward distant worlds. Instead, there is a stress on the commercial and the political potential of cooperation in space, more mundane features perhaps, but of vital importance in European integration. Whatever romance or glamour others may want to claim for cooperation in space, then, ESA and its supporters prefer the practical benefits, such as "economic prosperity"; in the words of Helmut Kohl, "We must, above all, think in terms of the future, but without losing sight of what is realistic" (European Space Agency 1989: 22).

The public discourse about ESA insists that although working in space may involve fanciful visions of the future and "avant-garde technology," these dreams are held in check by the pragmatic necessities of cooperation. It is pragmatism, indeed, that helps to temper the wild enthusiasms and runaway dreams of scientists and engineers intent on trying to produce the impossible. Thus, while Reimar Lüst (former director general of ESA) describes space scientists as being "compelled by their curiosity to dream of placing ever more expensive instruments in space to investigate objects of which they have little or no knowledge" (Lüst 1987: 1), he also assures his listeners that these same scientists do not always "have their head in a cloud, or their eyes to a telescope" (ibid.: 2), but in fact are able to focus on the here and now as well.

ESA's public relations materials place particular stress on counteracting any suggestion that impractical dreaming might be motivating the work of ESA, an expensive agency to which member states contribute millions of dollars. For instance, a 1983 brochure (European Space Agency 1983: 2) proclaims that although "to many people, space research looks like an extremely expensive way of giving a very limited sector of the community new and exotic toys to play with," it in fact has resulted in "new benefits to mankind," including improvements in television, communications, and weather forecasting. Another public relations piece insists that "the year 2000 will usher in the era in which the man-in-the-street will have a direct link to Space, the era of the satellite at the service of everyone" (European Space Agency 1985: 34). This is a vision of the future, to be sure, but it is one that speaks to the practical everyday needs of "everyman."

Cooperation may be the pragmatic answer to a variety of concerns, but questions still arise: How is cooperation to be accomplished? What principles should guide it, and what shape should it take? The answer that ESA offers is "harmonization," specifically the way in which the organization consolidates and implements its policies on an executive level: "the concept of harmonisation" is the "coordination of efforts . . . , guaranteeing the interests of all parties while at the same time ensuring the efficient orientation of all efforts" (Longdon and Guyenne 1984: 151).

Harmonization is not simply another word for cooperation; it is a particular form of cooperation. It reflects the idea of "unity in diversity," which, as I discussed in chapter 1, is characteristic of the European ideology, not simply of the technological requirements of a space agency. Harmonization is thus not peculiar as a strategy and a philosophy to ESA; it characterizes as well the workings of the EC and other organizations committed to the cause of European integration in other domains. It is an explicit policy of the EC, according to C. Twitchett (1980: 65–66), where it refers to the coordinating of "technical and administrative legislation among the Member States . . . required for the common market to function effectively."

Such harmonizing is not, however, "designed to produce dull uniformity throughout the Community" (ibid.), as suggested by critics of the EC. Quite the contrary: harmonization privileges interdependence and a concurrent maintenance of sovereignty, as participants commit to a unified effort without ignoring the specific interests that distinguish them from one another. Harmonization offers the possibility for joint endeavor without authoritarian command; ESA and analogous supranational bodies accordingly take on the responsibility for "coordination," which is a passive kind of leadership.

Despite the linguistic roots of the word, harmonization is not about achieving a feeling of harmony, nor is it about producing solidarity, in the sense of an affective consensus. Harmonization is part of pragmatic cooperation, because it is about efficiency; harmonization makes cooperation efficient—that is, both economically streamlined and technically focused. It enables ESA to take the idiosyncratic characteristics of the participants (e.g., national space policies, commercial industry's bottom lines, and national political goals) and weave together their distinctive interests without eliminating their differences.

In so doing, harmonization gives European nations and industries the ability to compete with the superpowers that dominate the international domain of economics, and the space race in particular. Without the unity produced out of harmonization, individual European nation-states competing in the global arena find themselves absorbed into the methods, interests, and demands of the United States. In this way, they seem to lose their distinctiveness, which is especially odious to these nation-states which share the belief that "differentiation is superiority" (Herzfeld 1987: 60). Cooperation among European nations is thus, in this view, necessary to enable them to distinguish themselves from the United States not only in space, but in economics and politics more generally. Cooperation permits them to interact with the United States as equals, rather than as little brothers to a big brother.[4]

Europe escapes "unity" with the United States by insuring "unity" at a regional level among European states. These states in turn resist such "unity" by demanding that their individual interests be respected in the method and policy of harmonization. Cooperation through harmonization thus insures autonomy and unity at multiple levels; unity at the level of the superpowers is built on the abdication of autonomy at the lower level of nation-states, just as unity at the level of the nation-state is built on the abdication of regional autonomy within the state's borders.

In this sense, the pragmatic approach of harmonization in ESA resonates strongly with the organizing principle of "subsidiarity" now guiding institutional developments in the European Community. As Holmes (forthcoming) indicates, "the unusual utility of the principle [of subsidiarity] is that it fosters a conceptualization of political integration as simultaneously centralizing and decentralizing." The pragmatic rhetoric of cooperation is particularly powerful in this context not so much because it asserts any moral claim (moral arguments are conspicuously absent in the discourse surrounding ESA), but because it allows for the production of both unity and autonomy, permitting the realization of individual and collective interests simultaneously.

Both subsidiarity and harmonization represent pragmatic efforts to make real the agenda of "unity in diversity" that has marked statist political ideologies in Europe. This ideology, as I have indicated, commands the combining of interests without their eradication, at least at the borders where differences are permitted to persist (Herzfeld 1987). The idea of cooperation, whether in ESA or in the EC, reflects the power of this ideology, linking the moral imperative of peace to more overtly pragmatic goals. In other words, cooperation satisfies the dual requirements of (social and cultural) diversity and (political and economic) profit that European institutions are compelled to address. Harmonization is in this way critical to the achievement of cooperation in ESA because it is a way of making spontaneous space music out of independent instruments.

Cooperation as Context: The Organization of ESA

As I have argued here, ESA's role is not necessarily to direct all European ventures into space—indeed, this would be inimical to the method of cooperation in Europe. Instead, ESA's role is to coordinate the independent activities undertaken by the member states, helping to insure that these states avoid any overlap or redundancy in activity and goals. ESA also provides the opportunity for these states to participate in projects otherwise too large, expensive, or complex for a single nation to pursue on its own, by facilitating the integration of disparate technological and financial efforts.

As an agency dedicated to cooperation and serving as its facilitator, ESA offers neither a pure research environment as in a university, nor a for-profit research-and-development environment as in industry. Rather, it resembles a government laboratory, only in this case it is intergovernmental. As such, it becomes the arena in which cooperation can be realized: it provides and acts as a forum for interaction, a locus for joint work, a purveyor of technological and political integration. ESA is a mediator, operating through and between various concerns, and this has entailed a multitude of difficulties as well as possibilities—difficulties associated as much with the design of its operational and bureaucratic procedures and structures, not inconsequential components of its activity, as with the design of hardware and technical systems. In this section, I present an overview of those ESA structures and policies which reflect the demands of harmonization in a European context. I use the present tense to describe these structures where the situation remains the same, the past tense for cases that have changed.

The thirteen permanent member states of ESA are Austria, Belgium, Denmark, France, Germany, Ireland, Italy, the Netherlands, Norway, Spain, Sweden, Switzerland, and the United Kingdom; Finland is an associate member.[5] In order to participate, member states must agree to support the mandatory programs of the Agency: the Science Programme and the General Budget (which supports administrative costs and in-house technical assistance, such as quality control, safety assurance, data compilation and archiving, and so on). This apparently "authoritarian" requirement is tempered in practice, as any changes to the budgets of these two programs must be agreed on unanimously by the participants. The size of each state's obligatory contribution is based on the average income of the state. Accordingly, the larger and wealthier nations (e.g., France and Germany) contribute a larger percentage than do the smaller nations (e.g., Ireland and the Netherlands) to the mandatory program budget.

In addition to this support, member states may choose to provide funds to any optional ESA programs in which they wish to participate. These optional programs include a wide variety of space applications projects, including earth observation, Columbus (space station) platforms, telecommunications, and microgravity research. Only participating states—that is, those contributing financially—have a say in the policy and operations of these optional programs. The size of the contribution to optional programs is based not on the relative wealth of participating states but rather on the degree of interest. For instance, France has always been a leading player in the development of the heavy launcher Ariane, and in 1988 it contributed the majority of the funding for this optional program, whereas Austria, Ireland, and Norway had elected not to participate at all in this effort.

In 1988, ESA's budget (mandatory and optional programs combined) totaled 1,903 million AU (accounting unit). The largest single national contributor was France, at 21.5 percent of the budget, closely followed by Germany, at 17.4 percent. The smallest contributor was Ireland, at 0.1 percent of the overall budget. Approximately 30 percent of the total income came from other, unspecified sources. When it came to expenditure, the single largest program was that of Space

Transportation Systems at 32.3 percent of the budget. Funding for the mandatory programs (General Budget and science projects) constituted 21.5 percent of the total budget (Longdon 1989: 186).[6]

While ESA does carry out a certain amount of research and development in space technology and space science in its own facilities and with its own staff, much of the actual design and manufacturing work on space projects is carried out by and in European industry. It is ESA's industrial policy, in fact, that articulates the essence of its coordinating, rather than directing, role. In addition to promoting the competitiveness of European industry (particularly in the domain of aerospace technology), the goal of this policy is to insure that the member states contributing such significant financial support to the Agency receive a "fair return" on their investment; this policy is called, accordingly, the policy of *juste retour*. It stipulates that member states "expect to receive for their national industries contracts of a value roughly proportional to their original contributions" (European Space Agency 1983: 11). Accordingly, if France provides 50 percent of the budget for development of the Ariane 5 launcher, then French industry should receive 50 percent of the contracts from ESA to undertake the actual design and manufacture of this rocket.

ESA awards these contracts for technological research and development and for the manufacture of particular mission components to consortia of industrial firms, which are themselves international and technically diverse in scope. The Agency tries to allocate these contracts in such a way as to permit a diverse group of firms to work on cutting-edge technology, and so to gain experience and to perfect performance in commercially profitable areas. This complex process of allocation and remuneration is an expression of "harmonization," in that industrial and national interests are safeguarded and not usurped by ESA.

Although ESA awards its contracts through a process of competitive bidding by industry, this competition operates within the constraints imposed by the industrial policy itself, constraints that emphasize the mutual sharing—across states and across industrial firms—of technological benefits and financial rewards: "[the contract allocation policy] will continue to be one of maximum competition within the constraints imposed by circumstances [i.e., of cooperation] and the implementation of an industrial policy aimed at developing and protecting the space industry of small countries" (Longdon and Guyenne 1984: 214–15). This policy requires delicate coordination of a variety of conflicting interests; such conflicts require ESA to assess the trade-offs entailed by any decision regarding industrial contract allocation. For instance, sometimes the goal of geographical representation (i.e., juste retour), combined with the goal of improving European industry in general, can mean the selection of a subcontractor with less experience in a particular technical area over one whose work reputation is already well known, in an effort to give work to a state whose industries are underrepresented, or whose industries need to gain technological expertise.

The interests of the individual participants in ESA missions are protected and looked after not only in ESA's industrial policy; they are, in fact, incorporated into the ongoing running of the Agency itself. ESA is governed by the ESA Council, which comprises political delegations from each member state. The Council, meeting four times a year, is responsible for making policy decisions relating to the full scope of the Agency's activities, including the areas of science, technology, administration, as well as industrial and international relations. The Council is assisted in its decision-making capacity by a variety of Programme Boards pertaining to the specialized programs run by the Agency, and by four general committees: the Science Programme Committee, the Administrative and Finance Committee, the Industrial Policy Committee, and the International Relations Advisory Committee. These Boards and Committees also comprise political and specialist delegations from each member state.

The decisions of the Council and its Programme Boards and Committees are carried out by the staff of the Agency who make up the executive branch. At the head of the Executive is the director general, who is appointed to a four-year term by the Council.[7] He is assisted in his work by the heads of the nine directorates of ESA. The directorates reflect both the optional projects undertaken by ESA and those areas necessary to support all space activities (see table 2.1 for a list of these directorates). The Agency staff is an international staff, hired to execute Agency policy and activities, not to represent independent political or national interests. The technical staff members serve primarily (but by no means exclusively) as monitors and coordinators of progress and standardized development of technology in industry and university laboratories, rather than as producers of systems and hardware. ESA staff, at the end of 1988 totaling 1,851 (Longdon 1989: 209), are spread over a number of sites.[8]

That the work of ESA is carried out at a number of different sites, located in different member states, is another expression of the harmonizing that characterizes ESA's procedures and structure. Because of these institutions' multinational placement, no one nation can be identified as the leader of ESA; ESA's work demonstrably requires the in-

TABLE 2.1
Directorates of ESA

Directorate	*Responsibilities*
	Mandatory Programmes
Administration	General management oversight of Agency activities, including accounting and budget procedures, publications, public relations, international relations
ESRIN	Information and archive services, including computer networks and data compilation and provision
Technical Directorate	Head of TD is head of ESTEC; general engineering and technological support, including product assurance and safety, equipment testing, research and development
Operations	Head of Operations is head of ESOC; mission scenario development, ground segment design and monitoring, telemetry tracking
Science	Scientific and engineering support for ESA scientific missions; some in-house scientific research and instrument development
	Optional Programmes
Earth Observation and Microgravity	Technological development of missions such as Meteosat (a meteorological satellite), ERS-1 (a remote-sensing satellite to conduct global ocean monitoring), Space Sled (an experiment to study the effects of weightlessness, flown on Spacelab)
Space Station and Platforms	Technological development of Eureca (an unmanned, orbiting platform for carrying scientific experiments); also development of the Columbus Programme, the European contributions to the U.S.-led Space Station *Freedom* (this directorate merged with the Space Transportation Directorate in 1994)
Space Transportation Systems	Technological development of the Ariane heavy launcher (version #5 currently under development) and of the Hermes spaceplane (to be used in conjunction with the space station); this directorate merged with the Space Station Directorate in 1994
Telecommunications	Technological development of missions such as Marecs (ship-shore communications) and Olympus (multipurpose telecommunications, including direct broadcasting and teleconferencing)

TABLE 2.2
Primary ESA Sites

Site and Location	*Size of Staff*	*Activities*
Headquarters in Paris, France	200	Political, legal, and administrative oversight, including offices of the director general and international relations (with non-member states, including the EC and the United States)
European Space Research and Technology Centre (ESTEC) in Noordwijk, the, Netherlands	1,250	Home of the Technical Directorate, SSD, and environmental test centers; staff conduct basic research and development in technology, support technological development on ESA projects, and oversee product assurance and safety analysis
European Space Operations Centre (ESOC) in Darmstadt, Germany	300	Home of the Operations Directorate; staff oversees operations and maneuvers of all ESA satellites (e.g., satellite tracking and control), and develops mission flight and trajectory scenarios and ground segment systems
European Space Research Institute (ESRIN) in Frascati, Italy	100	Now the Information Service and Archive of ESA; staff is responsible for collecting ESA and mission documentation, developing and maintaining computer networks, compiling and disseminating remote-sensing data, and providing access to and support of scientific and technological databases

Note: These figures represent staffing levels in 1988. The numbers are only approximate and do not include the significant number of industrial contract staff working in residence at ESA sites; at ESTEC, ESOC, and ESRIN, contract staff numbered in the hundreds.

volvement of many nations. There are four primary sites: Headquarters in France, ESTEC in the Netherlands, ESOC in Germany, and ESRIN in Italy (for details regarding these, see table 2.2). But to carry out its work, ESA needs to draw on resources from around the world as well. Space, after all, knows no boundaries, and as the earth spins, people on the planet's surface have to cross boundaries in order to monitor, track, and collect the information being dispatched by space satellites.[9]

In this way, the division of labor in space activities occurs both in terms of the political and economic demands of nation-states and in terms of the practical and technical demands of technoscience and bureaucracy. Moreover, ESA must coordinate these diverse interests in both its external and its internal relations. To do so, it implements policies of harmonization in multiple domains, policies that reflect the distinctive European logic of unity in diversity, and attempt to respect independence while promoting integration. ESA applies these policies externally to nation-states and national industries, and translates them internally into regulations concerning the coordination of differentiated expertise and technological artifacts.

The Science Programme and Scientific Missions

In the first part of this chapter, I have described ESA as an agency constituted by and for cooperation among European nations in space activities. These space activities include a wide variety of projects, as I have indicated, but I am focusing in this book on projects carried out under the rubric of the Science Programme. In this section, I turn my attention to this program, and to the scientific missions carried out in its purview.

The Science Programme, as I indicated above, occupies an anomalous position within ESA: it is a mandatory program, the heart and soul of space activities; but no longer considered a priority, it holds the status of a poor relation to the optional applications activities. Space science began as the foundation stone for joint European efforts in space and now takes its place as only one stone among many. The current position of science in ESA came about in part at the behest of the member states; since 1971, the orientation of the Agency has turned increasingly to commercial and industrial concerns. Although European scientists continue to oversee the implementation of science policy through various advisory committees dedicated to science activities, they no longer have dominant influence over the larger policy issues confronting the Agency. They are simply one group of specialists among many, representing one set of interests in an agency with a wide range of activities. Indeed, between 1971 and 1985, a long period of stagnation set in for science in ESA. The Science Programme continued to coordinate scientific missions, and several were launched during this time, but the resources allocated to such missions were limited.[10]

Despite the decreasing interest in science activities on the part of national governments, science retained a critical place in the Agency, a position of symbolic significance. Science was, and is, largely perceived by those in ESA as the key element keeping nations working together: "it provides a central, unifying element which binds together the diverse interests of the various States in the service of the Agency in which are vested Europe's hopes for a future in space. It is a recognition of the importance of fundamental research in the foundation of that future" (European Space Agency 1983: 12). Why is this research such an important element of that foundation? According to Lüst (1987: 5), it is because science research stimulates "new ideas, fresh challenges, and technical innovation" in all areas of space activity. In this view, everyone *needs* the science program; it is, in Lüst's words, the "heart" of "European space" (ibid.).[11]

The fact that ESA and Europe could not give up science altogether was finally given concrete expression in 1985, when, in the new Long-Term Programme announced for ESA, the implementation of a revised program for science was highlighted. Called Horizon 2000, this program attempted to revitalize European space science by expanding and rationalizing the scope and range of science missions undertaken by ESA. It also entailed a corresponding increase in the Science Programme budget for the realization of its varied components—although even with the increase in the Science Programme budget for Horizon 2000, the percentage of funds for science in ESA's overall budget remains small.[12] In formulating this plan, furthermore, ESA involved scientists from all over Europe, from a wide variety of universities and national research institutes. ESA was at pains to demonstrate that its plans for science were not serving organizational interests at the expense of scientific ones. The program's framers were not a small coterie of ESA insiders (even if they were a small coterie of scientific insiders).

The Horizon 2000 plan, which still largely governs space science policy in ESA, calls for a number of missions that are large in scope, requiring not only a significant financial expenditure but also intensive engineering research and development, as some of the technologies required by the scientific mission concepts are still untried and even undreamt. The Programme includes four major "cornerstone" missions that necessitated the full commitment and forward planning of ESA, in part because the technologies involved are complex and because the scope of the missions entails significant expenditures. These "red" missions represent four of the major space science disciplines:

The Solar Heliospheric Observatory and Cluster combined mission (scheduled launch: 1995) to study topics in both solar physics (e.g., solar structure and properties of the solar environment) and plasma physics (e.g., properties of the magnetosphere and ionosophere)

The X-ray Multi-Mirror Mission (scheduled launch: 1999), an orbiting telescope for conducting astronomical investigations in the X-ray wavelength range

The Far Infrared and Submillimeter Telescope (scheduled launch: 2006), an orbiting telescope to conduct astronomical investigations in the submillimeter wavelengths of the infrared range

Rosetta (scheduled launch: 2003), a space probe for studying the properties of primordial bodies, specifically comets

The Horizon 2000 program balances these "red" missions with "yellow missions" and "blue missions," which respectively represent small- and medium-scale projects that ESA also undertakes. Although the cornerstones were identified early on, empty slots remained in the "blue" and "yellow" categories, to be filled over the coming years "according to the ESA science programme's standard procedure, i.e., through an open competitive selection" (European Space Agency 1985: 26). Blue mission cost-to-completion budgets for spacecraft only, excluding the cost of payload instruments, ranged significantly. At the low end was the 168-million-AU budget for Ulysses (launched in 1990 with a scheduled lifetime of 5 years), a space probe dedicated to the study of the interplanetary medium and solar wind. At the high end, 550 million AU were allocated for the Infrared Space Observatory (scheduled for launch in 1995 with a lifetime of 1.5 years), an orbiting space telescope to permit observations of astronomical and planetary phenomena in the infrared wavelength range.[13]

ESA is proud of the process it developed for choosing the blue missions, not least because of the way it demonstrates the Agency's commitment to cooperation and coordination, rather than to control:

> The fundamental and essential characteristic of this process is that it is entirely controlled by the scientific community itself, with the Agency providing the necessary technical support. . . . [Although it may look] like a complex and lengthy process, . . . it does have considerable advantages. In particular, it permits a full involvement of the scientific community at all stages and the successive competitive stages ensure that a good scientific project will be selected primarily on scientific grounds, with industrial and political considerations only coming into play at a secondary level. (Manno 1988: 9)

The selection process unfolds through four stages, from a first "Request for Mission Proposals," issued to the general scientific community by ESA, through to the "Final Selection," when the study (that is, a mission under consideration) becomes a full-fledged project, with complete commitment from ESA and the member states. This entire process can take as long as four years, with as many as five missions attaining the "Phase A," intensive finalist stage. As the quotation above suggests, throughout the process ESA works with members of the European scientific community who serve on various standing and temporary advisory committees, and with industrial firms, through contracts awarded to study various technological scenarios. The Agency gives final say to the community as it is represented by the Space Science Advisory Committee, which recommends which mission should be undertaken.

The missions undertaken as part of Horizon 2000 come under the rubric of the Science Programme, and their goals are explicitly geared toward scientific ends: to study and to discover new things about the earth environment, the sun, the solar system, distant stars, galaxies, quasars—in short, the universe. Although these are scientific projects, scientists are not the only participants; they form only one contingent in this process. Designing, developing, testing, and launching such missions requires the extensive involvement of engineers of all types (mechanical, thermal, electrical, systems), operations analysts, software specialists, technicians, administrators, and countless others.

There are two basic kinds of science missions, reflecting the differences between the two major disciplinary areas of space science research. Astronomers typically design observatory missions that give them a window into the universe, enabling them to study stars, galaxies, quasars, and other phenomena light-years away. These missions provide a telescope that orbits the earth and focuses electromagnetic radiation coming from these distant light sources. Space observatories are generally dedicated to one wavelength range, as the different subdisciplines in astronomy require different technology to detect the relevant radiation (i.e., astronomers tend to commit themselves to studying objects that emit light in particular frequencies: radio, submillimeter, infrared, optical, ultraviolet, X-ray, and gamma-ray). Orbiting telescopes, like their ground-based counterparts, are open for use by the scientific community; although individual scientists may be involved in designing and building particular telescopes and the accompanying instrumentation, they are not the sole users of the observatories. ESA invites proposals from all members of the scientific community, including scientists from outside Europe, who wish to use the telescope to conduct their research; these proposals are also evaluated competitively.

The other type of mission flown is the space probe. This type of mission is characteristic of planetary science, plasma physics, and cosmic ray science (the other disciplines included in space science). Space probes can orbit the earth, or they can fly to distant locations. Unlike telescopes, these probes are not designed to look far into the universe; instead, they make in situ measurements, measurements of the environment wherever the craft itself is located. For instance, if the spacecraft is orbiting the earth, in situ measurements might be made of the solar wind that streams past and around the earth. Alternatively, if the spacecraft travels to another solar system body, the probe might study the atmospheric conditions of the planet beneath it.

Both observatories and probes combine a spacecraft with a payload; the aggregate of spacecraft and payload is referred to as a satellite. ESA provides the spacecraft, contracting with industry to have it designed, built, and tested. Often referred to as a "bus," this spacecraft is simply a vehicle for carrying passengers; the passengers in this case are scientific instruments built not by ESA but by scientists from the wider community. That is, the payload is provided by instrument consortia or teams headed by principal investigators (known as PIs). These PIs are senior scientists based in universities, and not employees of ESA.

In the case of observatory missions, ESA provides the telescope in addition to the spacecraft. But the telescope alone cannot enable scientists to conduct their research. Simply focusing the light from distant sources, as telescopes do, is not enough; scientific instruments must be attached to the telescope. These instruments actually collect and translate the electromagnetic radiation that is focused by the telescope itself into information scientists can use. These instruments, constituting the scientific payload for observatory missions, include such devices as spectrographs, photometers, and cameras.

In space probes, ESA builds a spacecraft that is really nothing more than a vehicle for scientific instrumentation. These missions have in the past contained no major, single "facility" (such as a telescope) open for use to the community. They are often referred to as "PI missions," because the PIs and their teams are initially the exclusive recipients of data gathered by the instruments they have provided. These data become available to the wider community at a later date; after one year, the scientists must turn them over to a central archive (e.g., magnetic tapes of data are stored at ESA and are available to anyone who would like to make use of them). Instruments designed to measure in situ properties include magnetometers, imagers, and analyzers of various sorts; these examine a wide variety of phenomena, including plasmas, cosmic rays, magnetic fields, planetary surface features, and atmospheric chemical composition.

Spacecraft development takes place in industrial firms, through contracts allocated and funded by ESA. ESA's in-house Project Team of engineers (headed by the project manager) oversees this development, monitoring progress and assuring that specifications are met, schedules adhered to, and quality standardized. The payload, on the other hand, is funded by national governments. These governments provide funds to the mission *directly*, by giving grants to the PIs and Co-Is (that is, coinvestigators) at their home institutions. These grants are made in addition to the member states' annual contributions to ESA's General Budget. Each instrument, as indicated above, is built by a consortium headed by the PI, who gathers together scientific colleagues and technical staff, often located in a variety of countries, to design, develop, manufacture, and ultimately make use of the instrument. Although the majority of PIs and participants are from ESA member states, this is not an absolute requirement. Some missions have had PIs from the United States and participating Co-Is from other countries, including Japan. Depending on the size and complexity of the instrument being built, the actual development work may either be carried out in the laboratories of universities or research centers, or be handed over to industry for implementation. Contracts between PIs and industry for the construction of scientific instrumentation are handled independently of ESA, since the funding is also separate from ESA. A typical mission payload will consist of several instruments, from as few as four (on observatories) to as many as eleven or twelve (on space probes). The cost-to-completion budget for the payload is significant; for instance, in the case of the Infrared Space Observatory, the budget for the four payload instruments runs to approximately 150 million AU. The fact that PIs are responsible for overseeing such major expenditures (each instrument running between 30 and 60 million AU) from independent sources gives them status nearly commensurate with that of ESA project managers in their negotiations over mission schedules and specifications.

Once a mission has been approved through the selection process, the project enters into a new series of stages that demarcate its progress from development through launch. Phase B is the initial, developmental phase, when the design of the various hardware and software systems is finalized. Changes are often made during this process as the entire mission design is evaluated and reevaluated. Following Phase B comes Phase C/D, the active production phase, when spacecraft components and systems are being constructed and tested in industry, and instruments in universities and research laboratories. Typically, models of different subsystems and sections will be constructed first so that different aspects of the spacecraft can be tested. For instance, a thermal model will contain the essential materials that must withstand the temperature variations to which the spacecraft will be subjected on launch and in flight, but this model will contain no working electronic parts and may not meet any other specifications. Other tests performed include vibration tests, vacuum tests (during which equipment is placed in a vacuum chamber to test its performance in spacelike conditions), and radiation tests. All of these are designed to simulate the total experience of spaceflight. The space environment is harsh and spacecraft are subjected to fluctuating extremes of vibration (launch being the most intense), heat and cold, and intense radiation, as craft pass in and out of radiation belts while orbiting earth, or if in flight in the solar system, through varying currents of the solar wind. This is the most intense period of development, as everyone races to meet schedule deadlines and put the pieces together. It culminates in Phase D, the launch; once launched, the project is considered in operation. About five years typically pass between the commencement of Phase B and the moment the satellite becomes operational.[14]

Depending on the type of mission, scientists may have to work fast to collect the information from satellites in orbit, or they may have to sit back and wait for some time before the payload instruments are turned on. Observatory missions become operational a few weeks or months after launch, during which time the satellite is put into its proper orbit, all spacecraft systems are checked out, and the instruments are turned on and tested. Such missions can last anywhere from eighteen months (the proposed lifetime, as indicated above, of the Infrared Space Observatory), to more than ten years (the fortuitous if unplanned lifetime of IUE, the International Ultraviolet Explorer, which at this writing continues to operate, having celebrated its tenth anniversary in 1989), to longer (as in the original plan for the Hubble Space Telescope, a joint ESA-NASA mission, which was designed to last fifteen years, with new payload instruments placed on the spacecraft at five-year intervals).

Space probes have different kinds of requirements. For example, in the case of Giotto, a probe launched in 1985 to study Halley's Comet, the instruments collected the relevant data during an encounter with the comet that lasted for several hours, after an eight-month journey. In the case of the Cassini-Huygens mission, a planetary mission currently under development, the scientists will have a much longer wait. This joint ESA-NASA mission to the planet Saturn (to be launched in 1997) must be able to last at least six years after launch, the time it takes to travel from earth to Saturn. Once it arrives at Saturn in 2002, the Huygens space probe (the ESA contribution, designed to study the moon Titan) will be released; its instruments will collect data during the approximately three hours of its descent, until it reaches the surface of the moon, where its continued survival is uncertain.

As I have suggested here, it is only when the satellites are actually flying that the scientific research begins in earnest. Now, with the payload instruments turned on, the data begin to come back to the scientists on earth, eagerly waiting to begin analysis and interpretation. As they are just getting down to work, the Project Team is disbanded and the industrial firms turn their attention to new projects. The cycle begins over again.

Conclusion

The scientific missions I have described above constitute the focus of daily work in ESA's Space Science Department. They are complex, not only in terms of the research questions they are designed to answer, but also technologically, financially, politically, and administratively. It is in these missions and for these missions that the cooperation to which ESA is dedicated must somehow be realized. For the individual participants, both scientists and engineers, who are engaged in the daily technical work of scientific mission design, development, and operations, the division of labor discussed in this chapter—a division that both constitutes and challenges ESA, as I have shown—faces them squarely every day as they negotiate the shifting and complex local terrain of multiple sites, multiple disciplines, multiple nations, multiple languages, multiple firms, and multiple projects. Despite this potentially chaotic and confusing array of components, they have managed to be successful in working together to develop and produce numerous missions to space.

How *do* people work together in this multilayered environment, and what is their experience of working together? Why is it that people can feel both frustrated about and yet proud of their participation in European cooperation in space science? These are questions of practice. Having here given some sense of the structure of cooperation and its attendant policies, rhetorics, and meanings, I now turn my attention to questions of practice. This practice of working together unfolds inside and in terms of these structures of cooperation. In the following chapters, I investigate the local environments where this work gets done. My aim is to understand both how people make sense of the structures that provide them with resources for action, and simultaneously how people produce those structures in the very process of working together. As will become clear, the logic and values of the structures of cooperation reappear in the dynamics of working together, refracted through the lens of practice.

Three

The Practice of Cooperation: Working Together on Space Science Missions

> ME: How do scientists and engineers manage to cooperate given their divergent perspectives?
>
> ENGINEER: Well, we support each other—but that's not cooperation, that's what we're here for.

As I SUGGESTED in the preceding chapter, ESA is offered as the proof that cooperation can work, that it can produce the unity and autonomy needed and expected by Europe. The question remains, however, how such cooperation can be produced out of the multitude of differences implicated in daily work on space science missions, differences arising from a division of labor that extends beyond the functional tasks of technoscience to include the demands of political and economic integration. Some high level administrators and political supporters of ESA suggest that the Agency's success at cooperation is a result of "political will," of nations agreeing to put their common interests ahead of their competitive ones (see European Space Agency 1985: 13). But the political will required to instantiate cooperation in treaties and organizations does not address the immediate, practical problems of day-to-day work, problems not only of how to get people down to work, but of how to get them down to work together.

Participants in space science missions themselves attempt to establish and maintain a shared logic of practice in a complex environment of competing structures and interests. In this practice, ESA provides them with a facilitating resource, not a unitary framework; participants are, after all, members of separate and distinct organizations, not united in or encompassed by the bureaucratic hierarchies of ESA. Indeed, no one bureau, department, agency, or person exercises absolute control over everyone else in the process of putting together space science missions. In this process, ESA is perhaps best described as a common thread, one that people follow, manipulate, and ravel as they improvise their practice of working together. In the end, participants in working together can depend only on each other, and on their mutual commitment to space science missions, to insure that the practice of cooperation remains both lively and productive.

Because of this interdependence of participants, people close to the technical and professional work of the Agency suggest that ESA's success at cooperation has more to do with the ordinary practices of people doing their jobs, and doing them well, than it has to do with political will. An ESA administrator involved in coordinating ESA's international relations, when talking to me about the secret of ESA's success, emphasized that "it has to do with the nature of what we're doing—we recruit engineers, scientists, lawyers, and not civil servants." Successful cooperation, in this view, depends neither on the political agreements of nations nor on particular bureaucratic structures, but on professionals committing to a focus on technical and specialized tasks. The secret, that is, lies in the routine, yet partly magical, process of "working together," the ethnographic contours of which are the focus of this chapter.

Cooperation versus Working Together

The scientists and engineers I talked to always drew a distinction between the European cooperation manifested by ESA and their own daily work. For instance, in response to my questions about the success of cooperation in ESA, one ESTEC engineer told me that "these concepts are too vague and not important to the work being done." Indeed, except when talking to me about such matters, participating scientists and engineers never used the word *cooperation* in their common, everyday language. One SSD scientist explained to me that this was because cooperation is, in a way, "assumed," so people never have to speak about it, or refer to it in any way; an SSD scientist colleague of his, in a different conversation with me, agreed: "We don't need to say it [cooperation] because we're working on the same project."

From the perspective of participants, an articulated "cooperation" refers explicitly to the political machinations of nations, industries, and agencies as these institutions search for ways to pool resources in order to implement large-scale projects in space to the benefit of all and each. When such institutions sign agreements and contracts, they establish "cooperation" on a particular mission by putting in place bureaucratic procedures and structures that will bring together a diversity of actors in concerted effort. This kind of cooperation exists, however, only on paper, in the rules and regulations defining a project, in the lists of participants who are either assigned to work on a project or who volunteer to do so. These structures do not specify practices—that is left to the people on the ground to figure out and implement. During a meeting of a group of scientists working on a fledgling mission, I heard this acknowledged explicitly when one scientist said that "the collaboration exists but not the interfaces"; he was referring to the fact that there were lists of people and institutes committed to participation, but how they would actually work together had not yet been figured out. Participants thus did not associate the phrase "European cooperation" with their own local practices; instead, this was something far from their immediate concerns, because it was not itself the substance of work in technoscience. From the participants' perspective, "cooperation" did not describe or refer to what everyone was actually doing; it was instead the framework within which people had to work things out together: "Cooperation *really* means doing the work together," according to a young SSD scientist. It is the process of working together that gives life to the distant and silent structures of cooperation.

Working together is the practice of daily life, and as such it is taken for granted: it is just how things are, and so cannot be codified, reified, or otherwise established as a concrete entity for examination—except when an inquiring ethnographer might momentarily call attention to it. Working together is the daily routine, with all its problems and possibilities. It is an ongoing activity, noisy and dynamic, and rife with ambiguity, as people negotiate the differences that are the substance of their practices, differences emerging directly from the division of labor that provides the material and the opportunity for their work. Contradictions are endemic to the process; people talked about their work as involving both conflict and harmony, as being simultaneously voluntary and coerced, and as being both exciting and boring. This is one reason that cooperation seemed so remote from local experience; although the rhetoric of European cooperation stressed sharing, unity, and harmony as clear achievements and unambiguous goals, in the daily flow of working together these seemed, to participants, to be ever receding even as they strived for them.

Instead of the clean and unproblematic unity of European integration, on the local level people perceived a process involving diverse, distinct individuals and groups that often seemed more divided than connected. Conflict, competition, tension, boundaries, distinctions, and equality were one side of working together; on the other were harmony, solidarity, congeniality, commonality, and hierarchy. When people talked about working together, about how they kept doing it despite or perhaps because of the differences that separated them, these contradictions were evident, even if sometimes only by implication:

> Everyone is aware that quarreling amongst ourselves doesn't make sense . . . [working on this project involves] some negotiating, and you could imagine this could lead to heated discussion, but it's been handled on a congenial basis. —*An SSD scientist*

We really wanted to work as a group; [this meant that] no one tried to single himself out. —*An SSD project scientist*

[In this mission] everyone's feathering his own nest, [so] sometimes you don't get the whole picture . . . but if our efforts become fragmented, we'll never achieve anything. —*An SSD scientist*

[In this project] all the people there are working together to build the same thing; . . . beyond all the conflicts, everyone wants to get a consensus to build [this satellite]. —*An academic scientist, member of a mission Science Team*

Why do they do their best? Because they feel themselves part of a team and they discharge themselves with honor . . . since this is part of their own making. . . . We're not fighting each other, we are friends; . . . the prime objective should be to do the mission. —*An ESTEC project engineer speaking of other engineers on the Project Team*

When we work on something, we work as one. —*An SSD engineer*

As the preceding quotations suggest, people emphasized the positive, harmonious, integrating aspect of the working-together dynamic, but they did so partly because they were responding to their experience of—and my observations of—the negative, conflictual, and dispersing aspects. In a way, participants seemed to be unhappy about their inability to maintain, without interruption, the positive dynamic of working together, even if they were resigned to, and indeed accepted the inevitability of, conflict. Yet despite their expressions of frustration, people would not have it any other way; they defended their right to be independent and argumentative as a way of demonstrating, for one thing, that they were not the lackeys of political cooperation, which was to them a coercive framework for action.

I was surprised to discover that cooperation could appear as, and be experienced as, something negative, something to be resisted, when its rhetoric emphasized the values and goals of pragmatism and unity, which seem, at least to a naive American observer, to be unproblematically positive qualities. I came to understand that one reason participants in working together denied their involvement in "cooperation" was that they perceived the structures it put in place as not only separate from but inimical to the goals of those who had to work in their spaces. From the perspective of participants, bureaucrats and politicians seemed actively to be imposing cooperative structures on them; scientists and engineers who were working on scientific missions then "inherited" these structures because they were involved in mission work, whether they liked them or not. By contrast, their daily work seemed to flow onward according to a logic distinct from that of those who attempted to rationalize and codify practices in hierarchical structures. Scientists and engineers were convinced that it was *they* who were responsible for the success of ESA, despite the bureaucracy and the politics.[1]

The "secret" behind the structures of cooperation I described in chapter 2 thus lies here, at the heart of the Agency's technical activities, in working together, where people interact and produce, despite conflict and through conflict. This is an always unfolding process, offering no local resolution to the contradictions inherent in the division of labor, contradictions that both motivate and define the work at hand. It is simultaneously true that without the (constraining) political economy of European cooperation in space science, there would be no context for working together in the first place; one SSD scientist explained to me that the constraints which cooperation through ESA imposes are inevitable and necessary. To him, ESA represented "the only game in town" when it came to working on complex space science missions: "I take the boundary conditions and I take pride in working within them."

The expression "boundary conditions" here represents a discursive, practical understanding of what I am calling "structure." When scientists refer to "boundary conditions," they are drawing an analogy between aspects of their technical, scientific work and the context of political cooperation in which they find themselves working. "Boundary conditions," in scientific jargon, refers to the specific stipulations, physical and/or mathematical, that circumscribe the posing of a particular problem and its consequent solution. Thus, every problem has a particular set of boundary conditions that must be satisfied, but that also delimit the possible scope of application or meaning of the answer ultimately derived. When scientists extend the use of this phrase to their working lives, they are referring to a social, bureaucratic, or political problem that they must resolve or negotiate. They themselves have not created the parameters (boundary conditions) within which they must act, nor do they have the power to change them, even if they can make use of them in solving the problem.

The "boundary conditions" to which the scientist was referring in the quotation above were those established by ESA when it set up the particular mission on which he was working. These conditions or parameters included the bureaucratic organization and the composition of the various groups and teams involved in the process. The scientist not only acknowledged these parameters but accepted the fact that his work involved taking all of these contradictory, conflicting ingredients and resolving them in practice, as he endeavored to protect the scientific interests of the mission. He had to do this whether or not he liked the cooperative structures he worked in. This quotation thus reveals one way in which some participants explicitly recognize that social actors make use of structures which also constrain them, both accommodating to and resisting them. It is indeed through the negotiation of such structures of cooperation that the scientists and engineers carrying out their daily work engage in the practices of working together, and it is through this negotiation that such structures or boundary conditions are themselves produced (a point to which I will return in chapter 7).

The following sections of this chapter examine some of the elements constituting the daily work that goes on in terms of these and other "boundary conditions." That is, here I take up the ethnographic details of working together by focusing on where participants were working, who was doing the work, and what, exactly, it was that they were doing.

I begin with a consideration of ESTEC's physical plant and, more specifically, the environment of SSD. I treat the "where" not only to give some "local color" but because, as Traweek (1988) argues, the offices, laboratories, cafeterias, and corridors of research-and-development work in technoscience affect and reflect the organization of work and the preoccupations of participants. By focusing here on ESTEC and SSD, however, I do not mean to privilege ESA's role in the complex process of working together. As I have emphasized repeatedly, these are not the only places where work on ESA missions is executed; there is, in fact, no single space that unites everyone in common effort, only multiple sites where people come together periodically to do their work (although ESTEC, indeed, has much in common with other office and laboratory complexes where work on ESA missions is carried out). I focus on ESTEC and SSD here simply because it is the site where I carried out my fieldwork and came to know most of those participants—scientists and engineers from in and out of ESA—with whom I talked during the course of my research.

Next, I turn to the "who," as I consider the involvement of people according to the two primary categories of difference (national affiliation and occupation; see chapter 1) that are salient in ESA missions. In addition to discussing aspects related to "nationality"—a local term—I review in some detail the categories of ESA staff as well as the categories of participants and groups external to ESA that were integrally involved in mission work. As I explained in chapter 1, both national and occupational categories were critical parts of practice; it was indeed in terms of these identities of difference that people carried out their work. In other words, both nationality and occupation provided significant resources for negotiation, conflict, and consensus in working together, as I show in subsequent chapters.

I follow this exploration of the dramatis personae with a focus on the plot; that is, I examine what it was that people were actually doing, day in and day out, when they were working together. (The drama itself is revealed in subsequent chapters.) Throughout the presentation of ethnographic detail, I stress the complexity of working together, a process that involves continual negotiation of the differences which accompany the division of labor. Daily work is, indeed, messy, and people are always trying to clean it up by finding and even imposing order and harmony where they can. Nonetheless, they cannot ever really succeed in this tidying, and, as I will suggest, in some ways they do not really want to—or at least they recognize that this is an unattainable dream, even if it is a dream worth having.

Where People Work: ESTEC and SSD Considered

ESA's public relations literature proclaims that ESTEC is "the European 'think-tank' for our future in space." The physical plant, located in Noordwijk, the Netherlands (about twenty kilometers north of The Hague), sprawls over thirty-five hectares alongside high coastal dunes covered with sand and green bushes. Just outside the fence surrounding the plant are characteristic features of the Dutch physical landscape, such as red brick buildings and canals; inside the fence are features characteristic of Western industrial concerns anywhere—office buildings, a building housing a library, conference area, and cafeteria (known as the canteen), and laboratory and testing facilities. The site is also beautifully landscaped, in keeping with the Dutch surroundings: the buildings and parking lots are situated along a roadway that meanders through the site, past a pond, green grassy slopes, ornamental bushes, and, in spring and summer, flowers. There is also, just inside the front gates, a modern sports and recreation facility for staff members and their families.[2]

The ESTEC site is, overall, a study in sharp contrasts, combining the serious with the playful, the functional with the meaningful, oppositions that make sense in the local context. At one extreme is the utilitarian office space and testing facility, sites designed for the serious work of instrument and satellite development; at the other is the whimsical canteen, a site intended for relaxation and play. These two modalities of working together coexist at ESTEC, both physically and practically, although participants' practices did not always map neatly onto the physical spaces.

For the participants who were my main group of informants (ESA scientists and engineers working at ESTEC), work took place primarily in the serious spaces of ESTEC—the laboratories, clean rooms, meeting rooms, and offices of the bare, functional, and rectilinear building that was ESTEC's original home. More specifically, these participants spent most of their time in the latter two sites, making only sporadic trips to laboratories or testing facilities when the need arose. At the time of my fieldwork, these offices were located almost exclusively in the original ESTEC office building (the new office towers opened only in July 1989), a long, five-story structure off which ran numerous connecting five-story wings. The exceptions were a few offices and computer rooms of SSD, located in one of the several one-story barrackslike buildings that had been erected between some of the wings of the original structure.[3]

Inside the main building, the overwhelming impression is of concrete blocks and straight lines, linoleum floors, long corridors lined with uniform office doors, a color scheme dominated by khaki, gray, brown, and white. Indeed, there is little to distinguish one corridor from another, one floor from another, even at times, one office from another; as a result, the interior sometimes appears as a bewildering maze, particularly for the novice staff member or visitor to the establishment. Some employees do make an effort to distinguish themselves in their work spaces, decorating walls with posters, photographs, cartoons, and blueprints related to their activities. But this does little to alleviate the building's blandness. With no character, no artfulness, it seems to represent itself simply as function, as if insisting that its denizens have time only for the serious business of making space satellites.

This impression is reinforced at the far end of the office complex, where the testing facilities are located. ESTEC houses one of the three main satellite test centers in Europe. Here, space hardware is tested in the rigorous and stressful conditions the equipment would experience when launched into space. The environmental test center at ESTEC included clean-room facilities (atmospherically controlled to limit contact with dust and other contaminants) for equipment development and checkout, and a variety of testing equipment that vibrated, heated up, and cooled down hardware. The Large Space Simulator, in which entire satellites could be mounted, mimicked the actual conditions in space, including those of vacuum, temperature, and light. Visitors could observe the activities in the test center from a series of windows that lined one wall of the facility, while staff members inside, clothed in special green lab coats, worked with equipment, assembling and disassembling hardware, studying computer printouts, and mounting tests.

The stark and unremitting interior landscape of the office areas and the serious and clean spaces of the testing facilities are in striking contrast with the physical presentation of two unusual and playful new buildings, designed by well-known Dutch architect Aldo van Eyck: a building housing the library, canteen, and conference center (which opened in July 1988, just prior to my arrival); and the office towers (which opened during the summer of 1989, just prior to my departure from ESTEC). The architect designed these two structures to fit together aesthetically; however, they bear no visual resemblance to the original buildings on the site.[4]

Constructed of wood, the new office towers have a starlike shape, with curved sides intersecting at many points. This design gives all the offices inside numerous windows and irregular shapes. Special furniture even had to be ordered to fit into odd corners and multisided rooms. In contrast with the bare corridors of the original building, interior staircases and other architectural features are painted in bright colors. Yet another "playful" feature is that some staircases do not go from top to bottom of the towers; some instead terminate midway, making it very easy for newcomers to get lost, although not, this time, because of relentless sameness, as in the original building, but because of a kind of ironic idiosyncrasy, a challenging of routine expectations.

This same playfulness marks the building complex that houses the library, canteen, and conference facilities. This building resembles a pavilion, with sloping pagoda-style roofs at various heights, the highest and most central one being the roof over the canteen, with the library and conference rooms appearing as smaller circles around the outside of the canteen. The conference section comprises three large meeting rooms, one with booths provided for simultaneous interpreters, each complete with all the latest technological amenities. These rooms have high ceilings and tall windows, letting in light that shines softly against wood paneling and wood furniture. The visual warmth of these surroundings is complemented by the brightness of the canteen; indeed, the canteen provides the most unusual physical environment in all of ESTEC. It is a huge space, large enough to serve and seat five hundred people, yet it is carved up by shoulder-level wood partitions to allow for more intimate spaces. The ceiling and long-windowed walls are wood, and huge iron girders painted in bold colors (yellow, purple, red, green) arch across the space. With large white spherical lamps suspended at different heights, the canteen resembles a festival tent hung with Chinese lanterns.

These new buildings were the source of much debate among staff members during my time at ESTEC. They provided, among other things, a way for individuals to articulate their resistance to the structure of ESA in which they found themselves working, often, it seemed, against their will. For instance, some people complained about the new office towers, giving voice to their distrust of the design's playfulness. They did not like the oddly shaped rooms or the idiosyncratic staircases. These features seemed to them somehow "inappropriate" for a place like ESTEC, in their view impeding rather than facilitating the serious work that had to go on in these offices. To many technical staff members, the design's inappropriateness was further proof of the indifference of ESA management (with whom such decisions rested) to the demands of technical and scientific work, an indifference people expected and complained about generally.

The canteen was not the object of such uniform derision, however. People had been almost unanimous in their dislike for the old canteen, located in cramped quarters on the top floor of the original office building, and found the spaciousness of the new canteen a welcome change. Even so, the playfulness of this space also jarred some sensibilities; accordingly, the canteen, too, provided many occasions for people to complain about ESA, an organization with administrative and political concerns distinct from the technical preoccupations of participants. For instance, a recurring "technical" difficulty with the operation of the new canteen was a frequent target of joking commentary. The conveyor belt designed to carry trays with dirty dishes from the eating space back to the kitchen routinely broke down during the first year of the canteen's operation. As a result, people would end up stacking their trays and dishes on the floor all around the conveyor belt, which was right in the middle of the room, across from the salad bar. My lunch partners regularly remarked on this disarray, ironically commenting on ESA's inability to get this "complex, technical machine" to operate properly, or else shaking their heads in dismay at the messy and bumbling image that ESA was reflecting to numerous official visitors. The implication was always that if they (the scientists, engineers, and technicians) had been in charge of the technical design, problems like this one would not exist. In this way, participants distinguished themselves from ESA management, suggesting that in their view, bureaucrats were ill-equipped to run a high-tech organization which needed instead technocrats like themselves.

SSD

Although I wandered throughout ESTEC during my time there—visiting engineers in the offices of the Technical Directorate or the Scientific Projects Department, attending meetings, observing instrument testing, and eating lunch—I spent most of my time at ESTEC in the Space

Science Department, and came to know its spaces and its routines best of all. For this reason, I offer here a picture of the physical layout of SSD as it was during my fieldwork (1988–1989).

ESTEC is run by the head of the Technical Directorate, and except for the two departments of the Science Directorate (SSD and the Scientific Projects Department), most of the departments at ESTEC focus on the wide range of engineering tasks and problems that are the responsibility of this Directorate.[5] SSD is anomalous in this respect, home as it is to scientists. Despite this difference, however, the physical space of SSD looked just the same as that elsewhere in the building. SSD was dominated by office space, although there were also a few laboratories dedicated to the design and testing of scientific instrumentation. These offices and laboratories were located primarily along corridor 16, the "spine" of the original office building, and were spread over the first, second, and third stories; in this area, they clustered according to the three administrative divisions of SSD: Planetary and Space Science, Solar and Heliospheric Science, and Astrophysics. On the ground floor, along corridor 16, were located the laboratories and computer facilities of the Planetary Division, as well as several of the Astrophysics Division laboratories. On the second floor, down a long stretch of corridor 16, were a few additional Astrophysics laboratories and the offices of most of the Astrophysics Division staff, the Planetary Division staff, and the head of the Department. Next to the department head's office was that housing the technicians devoted to mechanical and computer-aided design for the entire Department. On the floor above were located the offices and laboratories of the Solar Division. As I have mentioned, some scientific staff of the Astrophysics Division were also located in two of the barracks buildings.[6]

Most of the space in SSD, as suggested here, was devoted to the work of particular divisions. Even informal social gatherings took place in division-focused rather than department-wide spaces. In the Solar Division, the coffee machine was located in the division's main electronics laboratory, and it was here that coffee breaks and impromptu parties were held, amid the clutter of machinery, electronic components, and computers. The Planetary Division had no such space; their coffee machine was located in a small room filled to capacity with machines: two laser printers, a mechanical drawing printer, and a photocopying machine. The Astrophysics Division combined their coffee space with their "Preprint Library," which contained scientific preprints and journals as well as ESA documents and official papers. This room also held three computer terminals, two laser printers, a graph printer, a label printer, and a fax machine. The only shared SSD space in this general area of the ESTEC building was the SSD

Library, a small room lined with bookshelves, without space for reading or working. The library stocked research journals of various scientific disciplines (such as the *European Journal of Astronomy*), journals of general scientific interest (such as *Science*), and some policy and organizational material relating to ESA.

The department staff offices, like others throughout ESTEC (exceptions were the offices of senior management) were cold and rectangular, with blind-covered windows on one side. Almost all offices were filled to capacity, either with people (temporary research fellows in SSD shared office space) or paper, as metal book shelves and cabinets filled all available space. Nearly everyone had a computer terminal to work at—these were connected to ESTEC's mainframe IBM and VAX machines—and some had a personal computer as well. Although the physical spaces manifested this much uniformity, as I suggested above people in SSD, as in ESTEC more generally, attempted to alleviate the blandness by decorating their office walls and doors. The corridor walls were also "decorated"; each SSD division had a public bulletin board located along the corridor near its offices. These bulletin boards seemed to be sites for information dissemination; they were typically covered with posters advertising conferences, notices announcing internal seminars or meetings, memoranda from management, and occasionally notices for scientific positions outside of ESTEC. Yet I rarely saw people reading these bulletin boards; indeed, they served less as information sources (people relied on staff meetings, word of mouth, or electronic mail and electronic bulletin boards for their information) than as identity sources, indicating what kind of work was going on behind a particular cluster of doors, and hence what kind of people inhabited those otherwise indistinguishable spaces.[7]

Who Does the Work: Citizens, Scientists, and Engineers

As I have suggested, the division of labor in European cooperation in space science means that it takes many different kinds of people to put together a space science mission. Participants on ESA scientific missions are distinguished by, for instance, occupation, disciplinary specialty, institutional membership, national affiliation, and geographical location. In other words, working on every mission are people from and in many different countries; working on every mission are ESA and academic scientists, ESA and industry engineers, and technicians, administrators, politicians, civil servants, and secretaries. I focus here on what I identified in chapter 1 as the two most salient categories of difference: national affiliation and occupation.

Diversity of National Affiliation

National affiliation was recognized by participants in "nationality," a complex term that included dimensions of both social identity and membership in a political body. In some respects, identifying differences by "nationality" is straightforward: participants in working together are citizens of a variety of nation-states, including but not limited to those nations which are members of ESA. The structures of political-economic cooperation in ESA establish this aspect of "diversity" at the outset. ESA staff members are hired according to a system of geographical distribution, which insures that each member state has a representative contingent on the international staff roster. The current staff complement does not always reflect the contemporary ratios guiding recruitment, however, because these are always changing, and because they are always being manipulated by managers.

The staff distribution policy is analogous to that of the juste retour industrial policy outlined in the previous chapter. The juste retour policy also affects the composition of the field of participants in working together: it insures that external, industry engineers also come from a wide variety of states, since geographical distribution is a principle of contract allocation. Even in the selection of payload instruments national criteria are not omitted, despite the rhetoric proclaiming that these choices are made strictly according to "scientific criteria." One PI told me that "the expertise existed in [my country] to do everything" to build the payload instrument, but "for political and practical reasons" (for example, cost), he and his collaborators chose to include colleagues from two other European countries. Such pressures, official and unofficial, result in a field of participants that includes citizens of a multitude of countries. In the case of Science Teams and instrument consortia, participating scientists are citizens not only of European nation-states, but of the United States, Canada, Japan, and the countries of the former Soviet Union. Most industrial engineers, however, are citizens of European nation-states. This is because the ESA governing boards are loath to take the money contributed by its member states and give it to other international competitors. For instance, should members of a mission want to hire an American firm to produce a particular component or system, they must provide reams of documentation and technical argument proving that no European firm has the relevant expertise. This limits the number of non-European participating engineers.

Despite the fact that participants came from different nation-states, most people—but particularly scientists—discounted the significance of "nationality" in relation to their work activities. When describing the character of their daily tasks, participants focused invariably on the challenges posed by the occupational division of labor—how to get software specialists and scientists to see eye to eye, how to merge the work of distinct instrument groups, how to find consensus among engineers and scientists. The fact that these same participants came from different countries was not given any attention in work contexts, whether laboratory sessions or meetings. A senior ESTEC engineer explained this absence of attention, saying, "People recognize that although they were born into different nations, they are living and working together without a problem." Or as an ESA administrator put it, "Once we work for ESA for a while, we get used to the international environment, so we don't notice it."

This is not to say that nationality was never significant. Indeed, participants constantly talked about nationality, a dimension of experience and social identity that became particularly salient during the breaks in the working day when colleagues interacted in contexts of leisure rather than work—over meals, during coffee breaks, at social gatherings such as birthday parties, good-bye lunches, and Christmas dinners. These breaks were themselves part of the working-together process; thus, nationality was an important cultural category in working together.

When people spoke about nationality, they were referring specifically to the "mentality" and "culture" that citizens of particular nation-states exhibited and practiced ("culture" in this environment meant particularly clothing, cuisine, and custom; see Zabusky 1993a). This was a major source of the "diversity" that participants enjoyed and indeed prided themselves on. Other sources of personal identity, as I discussed in chapter 1, such as regional and ethnic differences within nations, were barely recognized and were deemed significant only to compatriots. For instance, when two German colleagues met for the first time, they might size each other up based on where they were from more locally within Germany; with other, non-German colleagues, however, this kind of distinction was largely irrelevant to their interactions.

Participants in working together entered into their collective activities with already well-developed notions of what constituted the various European nationalities represented by their colleagues. Such national stereotypes provided a means for instant ease of interaction when colleagues first began working together, because they organized the meaning of different nationalities in such a way as to tidy up expectations (and realizations) of clothing, cuisine, and customary behavior. Such stereotypes also carried with them a certain tension, however, since they inevitably involved a component of judgment, based on generalized notions that might or might not coincide with the actual clothing, cuisine, and customary behavior of particular individuals (see McDonald 1993 for a discussion of "stereotypes"). Nonetheless, stereotypes were the "common currency" (Herzfeld 1992) of daily interaction, as both other- and self-ascription. Indeed, people came to exaggerate certain "national" characteristics of their own when confronted by the diversity of nationalities embodied in their colleagues; such exaggeration appeared to be a means of individuating themselves, rather than effacing themselves, in certain collective contexts.

The orderliness provided by stereotyping was revealed in the shared understandings of what constituted the various categories of nationality. Everyone was able, for instance, to speak in largely similar terms about what characterized the "culture" of Germans, French, Italians, and so on: the foods they ate (for example, sausage, nouvelle cuisine, or pasta); the clothes they wore (formal suits, well-tailored jackets, or informal attire); customary behaviors (such as facility with the English language, or punctuality, or rule following and rule breaking); and more abstract features of their "mentality" (the Germans were "formal" or "rigid," while the French were "Cartesian," and the Italians were "informal" or "flexible").[8] Such shared distinctions were useful not only because they helped people understand how to interact with others; they also defined people's identities and helped to determine relationships of social distance and intimate connection. For this reason, people often seemed to celebrate national distinctions in moments of leisure, even though in work contexts these same distinctions were rendered irrelevant (Zabusky 1993a).

Diversity of Occupation

Although nationality was an ever-present feature, when it came to working together, said one ESTEC engineer of his co-workers, "one thousand different people are one thousand individuals, and it doesn't really matter where they come from; what matters is who they are." "Who they are" from his, and others', perspective, had much more to do with what work people were doing—their occupation, their discipline, or the tasks that engaged them.

ESA STAFF

ESA scientific missions are administered by the Science Programme. The head of this Directorate is based in Paris, in the ESA Headquarters, where he works along with a small senior staff including an executive secretary for science, a secretary for solar system science, and a secretary for astronomy. These members of the "executive branch" represent the Agency's scientific policy to academic scientists, both in Europe and elsewhere. The head of the Directorate also oversees the two departments based at ESTEC that together are responsible for carrying out the scientific and technical work of scientific missions: the Scientific Projects Department (Projects) and the Space Science Department (SSD).

The Projects Department is responsible for the technological development of those satellites which have been approved as part of the Scientific Programme's slate of missions; it is organized according to dedicated Project Teams, which means that every team of engineers focuses exclusively on a single mission. These teams oversee the development of technology in industry, both elaborating on and modifying specifications, and assuring quality and schedule control. Some members of these mission teams work with academic scientists and university engineers in developing software and data systems to link the payload instruments to the spacecraft. These Project Teams are established when a mission has been approved, and disband again when the mission has been launched and becomes operational. In addition to the multiple Project Teams, there is a division—the Future Projects Division—responsible for overseeing the development of studies, those missions which are still in the competitive phase of the selection process. The engineers in Future Projects work on such development not in the company of dedicated teams devoted to the missions in question, but with staff engineers from the Technical Directorate, a portion of whose time has been allocated to the missions under study.[9]

The engineers in ESA are identified in the division of labor by disciplinary expertise; distinct specializations are represented by different job titles on Project Teams or by different departmental affiliations: electrical subsystem engineers versus thermal engineers; Mechanical Engineering Department versus Quality Control. Some engineers may have specific training and competence in the field of aerospace technology; others may have specific training and competence instead in such disciplines as electrical engineering, mechanical engineering, or mathematics and computer science. These diverse areas of expertise are brought to bear on a common problem in the context of a Project Team, which is a structure of cooperation embedded inside the overarching structure of ESA. The fact that engineers have such distinct expertise means that they often find fellow team members' work distant from their own, even though they all share a common affiliation, which sometimes even means sharing office space. (The division of labor according to expertise or specialization holds as well for engineers in industry.)

In contrast to the Projects Department, SSD is dominated by scientists, rather than engineers, although scientists are not the only occupational category represented there—in addition to scientists, there are also engineers, software specialists, technicians, and secretaries. The staff, as indicated above, are spread out among three divisions that ostensibly represent scientific disciplinary boundaries. Each is administered by a division head, in each case a scientist who has been with SSD since the earliest days of the Agency.

The engineers in SSD have responsibility for overseeing work on the smaller experiments with which staff scientists get involved. This responsibility includes not only working with their hands in the laboratory, but also monitoring work being done either in outside university laboratories or in industrial firms, under contracts with SSD, that do the actual instrument development. Much of the basic engineering and hands-on technical work in SSD is done by the technicians, who represent a distinct category of worker. Technicians, unlike the engineers, are craft workers rather than professionals in the strict sense of that term; this reflects a European education system that provides extensive apprenticeship opportunities for skilled workers, although there are differences in categorical ascription between Continental engineers and British engineers. Like the engineers, the SSD technicians are also involved in the development of instrumentation and software. Some develop specific instrument configurations and test them; for instance, one technician is dedicated to work on infrared astronomy instruments such as Fabry-Perots, a particular kind of detector device. Others work on more general tasks, such as making electronics circuit breadboards or doing mechanical design (nowadays done as computer-aided design drawings) for a wide variety of missions in a cross section of scientific disciplines. The SSD laboratories are in fact primarily the domain of the engineers and technicians, who spend the majority of their time in them, working with computers and hardware.[10]

But it is the scientists who make SSD special. Indeed, although other departments at ESTEC have engineers, technicians, and secretaries, SSD is the only department at ESTEC, and in the Agency at large, that contains scientists.[11] SSD thus provides a small island of science in the midst not only of the larger enterprise of ESA itself, but in the institute that is its home, ESTEC. In 1989, out of the sixteen hundred ESA and contract staff in residence at ESTEC, only ninety were in SSD, and of these approximately fifty were scientists (including both permanent and temporary staff).

SSD scientists come in a variety of shapes and sizes, although to outsiders they appear to be significantly more alike than different. Among the younger scientists are the temporary research fellows, postdoctoral research associates hired for one or two years. Although many of these research fellows pursue their own independent research projects, some also get involved in ESA science missions, working on some small aspect related to their own research. Moreover, because of their temporary terms of appointment, research fellows are always coming and going; during the time of my stay in SSD, about six departed, and another four or five arrived.

Another coterie of temporary scientists is that of the "supernumerary" staff scientists. Supernumerary staff scientists are hired by ESA to participate in a particular mission, and because their contracts are restricted to the duration of that mission, they do not become part of the regular scientific staff although their stay with ESA can last several years. In the past, it has been primarily astronomy missions that have required the services of supernumerary staff, who served as the observatory scientific support staff while the satellite was operational. Some of these supernumerary staff carry out their work at various "outstations" that ESA has established for the operations of these astronomical satellites; others are on-site at ESTEC. These supernumerary staff scientists, except for their "temporary" status, resemble the regular staff scientists in every respect in terms of their work responsibilities.

"Regular" staff scientists are hired on four-year contracts, which can be renewed for an additional six-year term. At that time, the scientist is either let go or awarded an indefinite contract.[12] In theory, staff scientists are to divide their time fifty-fifty between mission- and ESA-related work and their own research and data analysis activities, although in practice the demands of mission support work often curtail the amount of time a staff scientist can devote to independent research. When acting as study or project scientists, SSD staff scientists provide the link between academic scientists and ESA engineers. Staff scientists also, from time to time, serve as Co-Is on smaller experiments, instruments that will be flown on missions run by other agencies. In addition to such mission work, staff scientists may also be involved in some in-house instrument development work.

SSD scientists are distinguished not only by the missions or instruments they work on but, even more significantly, by the specific disciplines or subdisciplines on which their research focuses. In SSD, these disciplines are represented by the different divisions, and then within the divisions by the different areas of research interest (e.g., astronomers versus planetary scientists versus solar-terrestrial scientists; X-ray versus infrared astronomers, or cosmic ray physicists versus helioseismologists). Scientists in different areas of research share a common ground only in the most general sense—I once even heard an argument between two SSD scientists from different divisions about the definition of "what every physicist should know." Scientists are further distinguished by the kind of work that they do; experimentalists work with hardware, designing instruments, while theoreticians work with data and hypotheses. Instrument people and data people find themselves far apart in their needs and interests. (This same differentiation of expertise or specialization holds as well for academic scientists.)

The main staff contingent in both the Projects Department and SSD—identified by participants as, respectively, engineers and scientists—is characterized, thus, by internal diversity, and insiders talked often about what divided them from each other, evaluating and defining their colleagues according to the particular subcategory in which they fell. To outsiders, however, these professional categories identified a homogeneous set of social actors; an engineer or a scientist was likely to discount the heterogeneity that members of the opposing group found particularly meaningful. In other words, to engineers, all scientists appeared to be more or less alike, even though for those considered "scientists" the differences between theoreticians and experimentalists, or between astronomers and plasma physicists, might be overwhelming.

The terms *scientist* and *engineer*, thus, represent "ideal types" in a Weberian sense; the terms are simplifying ones, making little or no reference, as I indicated, to differences among types of engineers or types of scientists, or to the complexities of practice. I use these terms to distinguish two primary social groups in part because participants themselves used them, although, needless to say, in practice everything was "fuzzy." For instance, although ESA made a rigid bureaucratic distinction between scientists and engineers—such that scientists were located exclusively inside SSD, and engineers primarily in the Projects Department and the Technical Directorate—there were in fact many people trained in scientific fields who held positions as "engineers" (there were no engineering-trained "scientists"). People's awareness of such complications did not change the terms of their discourse.

Context was also an important factor in the degree to which participants recognized the scientist/engineer divide as significant. For instance, this distinction was not particularly meaningful in contexts of scientific research or data analysis, contexts that tended to exclude the involvement of engineers altogether. And although the social distinction between scientists and engineers was remarked on by academic and industry participants alike, it was much sharper for those who worked for ESA. ESTEC staff members' discourse indeed highlighted the salience of this social divide, reflecting the constant daily interaction in ESTEC of social actors divided into these opposing moieties (in Zabusky 1992, I describe this distinction in some detail).

It was particularly in the collective contexts of working together, when the focus was on mission and instrument development rather than data analysis and theory, that the relationship between scientists and engineers appeared to participants as a critical component of the tasks at hand. Participants, particularly ESA staff members, understood that getting a space mission off the ground meant getting scientists and engineers to work together; getting them to do so was, however, no easy task, despite the fact that the bureaucratic structures of cooperation often specified lines of command and a rational organization of labor tasks. In designing instruments and scientific missions, participants recognized the interdependence of these social groups, sometimes arguing that scientists and engineers needed each other to produce high-quality, high-functioning technology. In these moments, the relationship of scientists and engineers was often perceived as complementary, as befits social relationships defined by a division of labor. One research fellow explained the relationship this way: "The science is important in deciding what the instrument should do and how; the engineer comes up with the practical design." That is, "the engineers take the abstract concept and get down to details."

Nonetheless, the dominant concern of each social group led people to focus more on what separated them than on what linked them. Indeed, the distinction between scientists and engineers was not only a social one; it reflected as well a cultural opposition, that between dreaming and pragmatism (Zabusky 1992). Scientists appeared as "dreamers," dealing with abstractions, "wild ideas," and "idiotic things." Engineers, on the other hand, were pragmatic, working with "two feet on the ground," bringing others "down to earth," and thinking about "trivial things such as budgets." To be a good scientist, said one SSD staff member, "you need a certain character, one of childish curiosity [that] you can never give up"; a good engineer, on the other hand, had to leave curiosity behind and focus on the "hard realities" of "compromise." Engineers had to calculate risks, taking into account the wide variety of constraints, whether technological, administrative, or financial, that confined and defined mission work; scientists, on the other hand, did not "have to take such responsibility," said one engineer. These "neat" categories of difference are decidedly more complicated than such statements make them appear.[13]

In working together, social actors find ways of converging despite and in terms of these occupational and disciplinary differences. Indeed, as I argue in subsequent chapters, differences between the perspectives and practices of scientists and engineers inform the dynamism of the process of working together.

MISSION PARTICIPANTS

In addition to the ESA staff in SSD and in the Projects Department, working together involves participants from universities and industrial firms, as I have indicated. On scientific missions, a key role is played by the Science Team. These teams are linked to ESA through SSD scientists who serve as study or project scientist for the mission. (Because the functions and activities of the study and project scientists are basically the same, for ease of reference I use the term *project scientist* to refer to both.) In this position, they act as the liaison between ESA and the "user community" of academic scientists who work outside of ESA's confines, translating the needs and desires of these "community" scientists into terms ESA managers and engineers can understand, and then carrying the policy stipulations and technical requirements of ESA back to that community. Project scientists in this way coordinate the development of the scientific payload; their job is to do this while safeguarding the scientific interests of the mission by working closely with the engineering team. As they shuttle back and forth between the Project Team of engineers and the academic scientists working on payload instrumentation, they act as "ambassadors" for science, taking turns in representing now scientists, now engineers, now the mission, now ESA (Zabusky 1992).

In acting as coordinator of payload development, the project scientist's primary responsibility is to work with the mission's Science Team. As indicated, the scientists on the Science Team are not part of ESA; they are members instead of national university departments and of national scientific research institutes (accordingly, I refer to all such external scientists as *academic scientists* in order to distinguish them from their ESA counterparts). These academic scientists are considered to be representatives and members of what participants call the "scientific community." In ESA parlance, moreover, the "scientific community" is commensurate with the "user community"; the term refers, in other words, to those people who will actually use the technology produced under the auspices of the Agency.

Among the academic scientists who are members of Science Teams are the PIs of the various instrument consortia that are providing the mission payload. There are also "mission scientists," academic scientists not involved in building particular instruments; these individuals, usually senior scientists with international reputations in their fields, represent the wider interests of the scientific community at large rather than the more specific interests of the teams building instruments.[14] These members of the Science Team do not work together in a daily, ongoing way, since they are based in various universities or institutes in different countries. To facilitate their common effort, participants come together at periodic intervals, three or four times a year typically, depending on the phase that the project has reached. At these meetings coordinated by the project scientist, they discuss the ongoing work, share information, and make decisions that pertain to the project's scientific dimension.

The Science Team is not, however, in charge of mission development, even though the instruments its members contribute will actually provide the data for the scientists to analyze. The academic scientists are not formally part of what ESA calls the *project*, even if they are part of the mission. Mission development is instead the province of the ESA Project Team, as I have indicated; from the perspective of the engineers, the Science Team acts only as an advisory body to their work. Even the project scientist, although an ESA staff member, is not considered to be part of the Project Team but only an adviser to it. This is made clear on the organizational chart of the Project Team, where the project scientist is attached to the Project Team only by a dotted line.

The Science Team does not appear anywhere at all on this organizational chart; it is the project scientist's role to interact with them on behalf of the engineers. There is, in fact, no formal organizational chart to represent the Science Team, only alphabetical lists of participants and instruments, partly because the Science Team is not administered by ESA, and partly because there is no hierarchical relationship to be depicted. All members of the Science Team have equal status; no one instrument is more important than another; no scientist has more say than another or any authority to command others to action.

The Project Team, unlike the Science Team, is organized hierarchically. The Team is led by a project manager, who oversees the technical dimensions of mission development. The project manager also controls all of the ESA funds devoted to the project at hand, particularly the development of the bus and any central "facility" such as a telescope. These funds are, moreover, committed not only to the development of hardware (through industry contracts and so on) but also to in-flight science operations and science development. The members of the Project Team work as a unit in-house, representing all the dimensions of the spacecraft and its connections to the payload. These engineers are not, for the most part, engaged in the actual design and manufacture of mission technology; this is the responsibility of industry to implement. Instead, they "run" the contracts with industry, monitoring the work done in their particular area of expertise, making sure that industry meets deadlines and fulfills the "specs" (specifications) stipulated by the ESA design, itself a product of collaboration between ESA engineers and industry engineers at an earlier phase of the project's lifetime. Their primary responsibility is thus to "interface" with external engineers, providing that coherence and connection to the overall mission which those working in isolated sites on separated components cannot achieve.[15]

The project manager, despite being in the powerful position of controlling the ESA funds, is nonetheless not the overall executive of a mission. Indeed, there is no such individual with total authority over mission decisions. Even the director of the Science Programme cannot mandate actions or secure funds that pertain to the payload. Thus, even though the Science Team is not in charge of the mission, neither is it entirely subordinate to the project. The external instrument teams bring their own money and their own forms of organization to bear on the collective problem that is the scientific satellite.

The PIs who lead these instrument teams are equal in status to the project manager, in part because they command and control significant financial resources (as indicated in chapter 2). Nonetheless, as is the case with the project scientist, they have no ultimate authority over the members of their own instrument teams or consortia. These consortia are themselves loose coalitions of diverse individuals from different institutes and disciplines. A single consortium dedicated to the design and manufacture of a particular instrument may include participants from several nations (not restricted to the member states), several institutes and/or departments, and a variety of disciplines (scientists, engineers, software specialists, and so on). Each group contributes a discrete component to the overall consortium, and in doing so brings its own money and its own form of organization, which the PI must coordinate. In this way, the structure of the instrument consortia viewed from the perspective of the PI mirrors the structure of the project viewed from the perspective of ESA.

The various requirements of missions as complex as these pose certain technical problems that cannot be solved by the members of the Project Team or Science Team alone. This leads to the establishment of various other groups dedicated to working on or working out such technical problems. These groups, generally called "Working Groups," are only partially formal; they are not mandated in the ESA bureaucratic structure, emerging instead in response to the needs of particular missions. For instance, for observatory missions, a Ground Segment Working Group may be set up to establish the procedures and systems that the satellite operations teams will use to communicate with the satellite, to organize the information that comes in to them from the community, and to distribute the data that are collected from the satellite. These groups are made up of members of instrument consortia as well as ESA staff members who have relevant expertise. These ESA participants may come from SSD, from the Project Team, or from other Directorates. For instance, crucial participants in project work are the staff of ESOC (ESA's telemetry center), who design and manage satellite operations systems, as well as study and develop flight scenarios.

What People Do: Technology and Talk in Working Together

Although the discovery of new things is the ultimate goal of a scientific mission, the working-together process finds its movement and energy not in such research plans—which involve only a small fraction of the participants—but in the negotiation of the technology that is the enabler of such (future) research. The preoccupation of all participants, then, was on the details of the technology. This focus on technology was as true for scientists as it was for engineers, even though many scientists denied this: "Engineers are interested in building things; scientists have a different aim, to discover new things." The "science" comes later, after the technological artifact is completed, when scientists make use of the technology to "discover new things." Indeed, as if to emphasize that it is after launch that science takes over, it is at this point that the Project Team dissolves away, leaving only scientists (and a small engineering crew involved with operations) to follow the satellite that previously consumed the attention and activity of so many diverse others.

During the development phases of mission work, engineers and scientists are engaged in a mutual struggle to produce functioning and successful technology. What this technology will look like is, however, not a foregone conclusion. On the one hand, it must be designed and developed to meet the mission's scientific goals. On the other hand, it must be designed and developed to be technologically feasible—that is, cost-effective, safe, and efficient. These twin constraints underlie much of the wrangling over technology design and implementation, and they typically get played out in terms of "science" or the "scientific requirements," and "engineering" or "technical specifications."

In this struggle over technology, scientists and engineers often find themselves on opposing sides, as they argue about which set of specifications or requirements should dictate design and testing. Scientists typically argue that it is the research goals that should dictate these technical parameters and the overall design of the mission concept ("or else we just stay home"). Engineers argue in turn that there are unavoidable technical constraints which, when taken into account, will insure both the feasibility and the safety of the mission, and they contrast their focus on such constraints with the "exotic requests" of scientists. In this tug-of-war, the SSD scientist takes on the role of the ambassador of science, shuttling between the two perspectives and the social actors who hold them.

What everyone is working toward, at every level, is the integration of technological components and systems. Integration (which is a technical term) means that diverse components are put together to form a unified working whole; such a whole is, in turn, integrated into yet another whole. The process resembles that depicted in the children's rhyme "The House That Jack Built": components become integrated as instruments; instruments become integrated as payload; the payload is integrated with the bus to become the satellite; the satellite is integrated with the fairing to become part of the rocket that is launched into space.[16]

On scientific missions, the most significant "integration" toward which participants are working is that of unifying the science payload and the spacecraft. As I discussed in chapter 2, missions are designed with the payload in mind, because it is the payload that enables scientists to acquire and analyze data; these instruments depend on the spacecraft—which is financed and provided by ESA, and developed and manufactured in industry—not only to get them into space, but for electrical power and protection from the harsh space environment.

From the scientists' perspective, although the payload rides in the spacecraft "bus," it is the bus that is, or should be, "only along for the ride," because it is there simply as an enabler and not as the mission's focus. For this reason, they are loath to limit or restrict the size and complexity of the payload, even in the face of budgetary restrictions. Scientists in fact frequently complained that engineers were overly focused on the spacecraft, at the expense of the payload. For instance, SSD scientists routinely argued with the project manager who, when there were mission cost overruns, often looked to the payload as an area that could be "descoped" or cut down; from his perspective, the payload represented "inessential" technology since the satellite did not depend on the payload to be able to fly. SSD scientists labeled this the "pointed brick" phenomenon, arguing that if such "descoping" were carried out to its logical extreme, the Agency would end up launching a satellite that could fly, and could be moved near or pointed toward an object to be analyzed, but that would carry on board only "a brick." That is, the satellite would have no instruments on board capable of collecting data.

The integration of payload with spacecraft is, as this example suggests, in some ways the most difficult part of the various integrating moments in the development of the technology. This is because it brings together technological artifacts designed to meet entirely different ends and developed by people with entirely different goals. This integration does not happen suddenly, in an instant, after the instruments and the spacecraft have been developed in total isolation from one another. Instead, it happens gradually, in the sense that the development of each major component proceeds in accordance with the overall requirements of a complete technological artifact.

For instance, payload instruments must be designed and developed in accordance with certain "technical specifications" issued by ESA, which has the responsibility for insuring the successful unification of parts into a working whole. These specifications dictate to the instrument builders such parameters as how large the instruments can be (both volume and weight) and how much energy the instruments can consume when they are operating. They also concern the instruments' radiation properties, indicating, for instance, what type of materials must be used in order to shield them from the radiation environment of space, as well as thermal properties, such as the temperature range the instruments must be able to withstand and the limit to how hot the instruments can get.

These specifications are geared to allow instruments to work side by side, so that one will not, for instance, consume more than its share of energy, since only a finite amount is available to the entire payload from the spacecraft systems. Similarly, no instrument can be allowed to grow too hot during operation or else it will affect the successful operation of a neighboring instrument, and no instrument can be allowed to grow too large or heavy during development or else it will encroach on the space and mass allocation given to another instrument. In sum, these specifications determine the uniform environment in which these diverse instruments will operate, even though the instruments themselves are developed in isolation from each other and designed to meet widely different ends.

The development of payload instruments is not, however, a one-way process, with ESA dictating all of these requirements out of hand.

The instruments themselves have requirements, in particular "scientific requirements," that must be met before they can carry out their intended observing or detecting tasks (*scientific requirements* refers not to abstract scientific problems but to the configuration of technology that will permit such problems to be solved). These requirements influence ESA engineers and must be taken into account when they develop the technical specifications—after all, all instruments do not have to be the same size or the same weight, nor do they all have to use the same amount of power, although they may all have to be adequately protected from the radiation of the space environment. Instruments vary, and the design of the payload environment must allow for the diverse requirements of these instruments. In the process of developing the technology, furthermore, ESA engineers modify the testing procedures that assess whether specifications have been met in accordance with such scientific requirements.

Integration, then, is a process of give-and-take, but it is nonetheless a process of finite duration. In the end, integration takes place; when all the various instruments are plugged into the spacecraft, process becomes product. The final integration of payload and spacecraft transforms these two radically different components into a single entity—the scientific satellite.[17]

Despite the focus on technology in mission development, daily activities, especially in ESTEC, often take place out of sight of the technology itself. For instance, most meetings are convened around a committee table, and reviews for evaluating satellite and system design are conducted not in laboratories but in conference rooms. The technology may be present during these gatherings in a to-scale model of the satellite, or perhaps only in the multitude of documents to which people refer, documents that reveal the inner workings of equipment and software in their detailed designs or in results from tests. Photographs or slides of hardware as it currently exists may be displayed during presentations as well. The technology that is under scrutiny during such meetings, however, is physically hidden from view, existing on shop floors in dispersed industrial centers or on laboratory benches in far-flung scientific institutes. And when a mission is still in its formative stages, the technology may exist only in drawings and computer-generated pictures, themselves beautiful and evocative, which suggest the physical shape that the artifacts will ultimately take.

If so much of the daily routine of scientists in SSD, engineers on the Project Team, and members of the Science Teams did not involve analyzing data or building technology, then what were they doing? To answer this question, I now turn to a consideration of people's actual activities when they were engaged in negotiating technology in working together. Take, for example, the case of the project scientist, whose responsibilities and activities resemble in their details those of the PIs and of other team and group leaders. Daily tasks included coordinating meetings with the Science Team, tracking the progress of payload instrument development, attending meetings of PI teams, attending meetings of the Project Team to keep track of progress in spacecraft systems development, representing the interests of the Science Team to the Project, and communicating the needs and requirements of the Project Team to the scientists. The project scientist participated in many meetings, sometimes as chair, and wrote reams of minutes; traveled constantly to keep in touch with the PIs and their teams spread out all over Europe, and, in joint missions, sometimes in the United States and the republics of the former Soviet Union as well; read a great deal of engineering documentation; created public relations material, not only writing and editing it but also doing layout and proofreading; and drafted numerous reports in many different forms for various constituencies (e.g., ESA Teams, ESA management, the ESA Council, the Science Programme Committee, academic scientists, and industry engineers).

In short, daily work consisted of talk, travel, and paperwork.

Talk

Talk is the central mechanism of working together: as one project engineer said to me, "We have to talk whether we like it or not." People talked to each other all the time, using a variety of modes of talk: they made telephone calls; sent regular mail, faxes, and electronic mail; exchanged memoranda; attended seminars; and participated in informal conversations, as two people put their heads together to solve a problem. Most especially, people went to meetings. Sometimes these meetings were as formal as design reviews, at other times, more routine, as in regular team meetings. In the following chapter, I discuss in more detail the role meetings play in working together—meetings were, in fact, particularly critical forms of "talk." In this section, I focus on the more linguistic dimension of talk, as I examine the way in which people worked together using, or trying to use, a common language.

In the European space science environment, talk was not and cannot be straightforward—as citizens of different countries, the participants in space science missions spoke different languages and indeed were trained in their professional specialties in different languages as well.

When working together, however, all members of the group used one language, no matter what their mother tongues. Although ESA has two official languages, French and English, and all staff are supposed to be able to converse in each, generally at meetings at ESTEC and elsewhere, English prevailed. This is because English has become, over the last forty years, the lingua franca of science and high technology in an international context. Some ESTEC scientists and engineers even explained to me that they had difficulty doing their technical work in a language other than English, including their native tongue, because they had become so accustomed to working in English over the years of their involvement with ESA.

The emphasis on English as a working language did not mean that other languages were never heard in work spaces. For instance, because ESTEC was in the Netherlands, many of the support staff were Dutch and spoke Dutch among themselves. And in the canteen, over lunch, people spoke their native languages when socializing with compatriots. But most participants insisted that the kind of technical work they did required the use of one language; as they saw it, in fact, this distinguished working together from the cooperation of politicians and administrators.

For instance, in meetings of the ESA Council, the preeminent political body governing the Agency, interpreters provided simultaneous translation of the proceedings. Some of the essential official documents—regarding funding commitments and long-range planning, for instance—were even translated into the language of each member state. This would be unthinkable in the case of technical engineering or scientific documents; most scientists and engineers considered such wrangling over language and the insistence on the use of multiple national languages to be indicative of cooperation, that political structure which was conceived and adhered to by agencies and nations. In contrast, when working together, these local participants treated language simply as a mechanism for communication, without attaching to it any symbolic significance.

Nonetheless, individuals' English-speaking abilities varied widely. Those with more exposure and experience using English spoke more idiomatically and fluently, so ESTEC staff members' daily communication in English was largely unproblematic, although newcomers to ESTEC sometimes had trouble with the routine flow of English conversation. Not every participant was a member of an international institute where English was the norm, however. Scientists and engineers who worked in national institutes or laboratories were usually accustomed to speaking in their national language in daily contexts; this meant that their interactions in the context of ESA mission work were the only opportunities they had for speaking English, and some found the pressure of speaking English constantly quite taxing.[18]

When I asked participants whether this kind of uneven linguistic ability ever posed problems for working together on technical topics, I was almost invariably assured that it was not a problem. Yes, some acknowledged, occasionally misunderstandings do occur, but they are inconsequential, and easily addressed. I in fact witnessed many instances of sudden linguistic breakdown, particularly during meetings, but such occasions were invariably incorporated into the flow of working together with ease. Typically a speaker might suddenly find him or herself unable to think of the appropriate technical word in English. In one meeting, this happened to a French scientist as she was explaining some tests being run on an instrument under development by her consortium. She paused and then uttered the word in French; another French-speaker in the meeting translated the word into English, and she continued. No one remarked on this brief pause, and it did not essentially interrupt the group's attention to the technical details at hand.

On another occasion, a French-speaker had difficulty following a rapid-fire conversation between two native speakers of English during a meeting. He spoke in French to another French colleague, signaling his lack of comprehension; a few of the French-speakers present then had a brief conversation in French to clarify the issues under discussion, and the original questioner returned to English in order to offer his comments on the topic to the group at large. Many non-French members of this group understood and spoke French as well, but the clarification was carried out only by those whose native tongue was French. If, however, it happened that someone became confused by a conversation and began to ask questions, with no compatriot available to translate or clarify, other participants would make the effort both to speak more slowly and to rephrase their remarks. Usually this corrective sufficed; such linguistic breaches were closed so quickly and easily that they disappeared into the flow of conversation. That is why "misunderstandings" could be so easily dismissed when I asked participants directly about the problems of working together in a common language that was not the native language of all involved.

In some respects, in fact, it was the lack of a common native language that facilitated the working-together process, as some of those involved acknowledged in both public and private contexts. For instance, at ESA's twenty-fifth Jubilee celebration, Professor van den Hulst, a Dutch scientist who had been active in the formation of ESA, asserted:

> One of the blessings for Europe is that we have many languages. People think of that as a disadvantage. . . . However, by and large, I think one should look at it as an asset. . . . I think that the biggest danger [in speaking English together] is that you all think you talk the same language, which of course the Americans think. In Europe, we know we don't talk the same language, so we make a much greater effort to understand and to explain what we really mean, and that is really an incentive to understand. . . . [A] real linguistic effort is required to try to reach understanding, and this adds a certain flexibility, it adds a certain shock-absorbing effect. . . . [T]hese different languages force you into a moment of reflection. (European Space Agency 1989: 38)

This notion—that people who share a common language might actually have greater difficulty in working together—was echoed by Valerie Hood, a member of the International Affairs Division of ESA (based at Paris Headquarters). In an interview published in the internal newspaper of ESA called *In Orbit*, she discussed the difficulties the nations of Latin America were having in formulating a regional space agency on the model of ESA. She indicated that "far from being a cement, the common languages, Spanish and Portuguese, seem to be more a handicap than a help; probably because the differences in opinions, and discords, are not softened by relying on a translation!"

Participants in working together with whom I spoke locally also shared these sentiments, explaining that when using a native language together, people could be insulted more easily, because there was not the "room for doubt" that using a language not your own allowed. This meant that conflicts could flare up to unmanageable proportions when people with the same language worked together, because there were no opportunities comparable to those in the situations I described above for people to take a moment to reflect on what was being discussed, and to revel in the differences of people and languages before returning to the common language and common purpose of their discussion.

It was not only national languages that could pose problems; so, too, could the technical and bureaucratic languages of science missions. One tendency that complicated matters was the proliferation of acronyms. One Science Team mission scientist compiled a dictionary of common acronyms used in mission work, which he distributed to the scientists and engineers on the Project Team. The distinct languages of different disciplines also challenged participants. One engineer working on mission software explained to me that meetings of the Project Team are "superficial, because we don't talk the same language. The mechanical engineers talk about loads and stresses, while software engineers talk about other things."

Travel

One of the overwhelming problems of participation in mission work was how to keep progress happening and the various groups together despite participants' geographic and bureaucratic separation. Even in the case of a local team whose members might have offices alongside one another in one institute, as at ESTEC or a national laboratory, members of these teams were still effectively isolated from one another both by the focused and specific nature of the work and by the physical setup, which separated people into private or semiprivate offices.

The solution was to keep talking, as I have indicated, and one way to facilitate this was to travel. In fact, people traveled constantly. For instance, SSD scientists were more often out of their offices than in. Sometimes they might be in ESTEC elsewhere, talking with technicians or engineers in other parts of the site. Often, they were not at ESTEC at all, but visiting other ESA establishments, or other scientific institutes, attending meetings or reviewing ongoing work (for instance, of instrument teams). A typical project scientist spent as much as 25 percent of his work time—that is, excluding holidays, vacation, and sick time—away from ESTEC. When a mission was nearing launch, the project scientist might spend weeks or even months away from home, participating in launch preparations at the launch site, or at the operations center. The case was similar for PIs who coordinated international instrument consortia. One PI told me he liked to organize his travel plans in such a way as to insure being in the office no less than two days a week; typically, he was away from his home institute for the other three days.

That travel was an onerous, if necessary, part of mission work became clear at meetings that brought people in from afar. For several minutes, as the meeting room slowly filled, people conversed informally with each other, and their conversation often concerned the trials and tribulations of travel: late planes, bad weather, lost luggage, missed connections, customs delays, uncomfortable hotel rooms, tasteless food. At the end of these meetings, when discussion turned to the planning of the next meeting, there was often heated debate about where it should be held. People argued about the time of year ("If it's winter, let's go south where it's warm, to Spain or Italy"), the expense of travel ("If we meet in Germany, I have to fly, but I'm over budget already"), the distance and length of time required to travel longer distances ("If we hold this workshop on Tenerife, we'll lose two business days in travel"), the other attractions of different locales ("In London, we can do great Christmas shopping, but in Marseilles, we could lie on the beach, but there's certainly nothing to do in Heidelberg"). Wherever they went, the fact was that travel was a part of life in working together on space science missions in Europe.

Paperwork

Paperwork was another means by which talk was facilitated in working together. Paperwork as such is characteristic of any large-scale bureaucracy, and ESA is no exception. Participants in working together spent an inordinate amount of time with paper, not only filling out administrative forms (e.g., requests for funds, personnel evaluations, travel expense forms, meeting scheduling forms), but more importantly writing and reading technical papers and reports. I am not referring here to scientific research papers, but to engineering documentation, meeting minutes, progress reports, and public relations pieces (e.g., technical pieces for dissemination to aerospace industry or to university scientists). Offices were full to the brim with paper, kept in filing cabinets and thick binders, or piled high on desks and bookshelves.

Documents were themselves the stuff of meetings, which in other respects were examples of talk. Indeed, meetings provided an occasion for the generation of documents—progress reports, analyses of problems, notes on policy changes, and so on. These documents, prepared by all members of the team or working group, were ideally completed before the start of a meeting so they could be circulated and reviewed, although this did not always occur in practice. Often during meetings, the sound of pages turning and papers rustling accompanied discussions and presentations as participants reviewed materials just received. And after a meeting had ended, the chair was responsible for preparing and circulating the meeting minutes, which included a list of "action" items, a recapitulation of tasks assigned to participants during the meeting. Among the meetings I attended, some even focused on the form and content of particular documents, rather than on technological or scientific details as such. Indeed, it might be said that paperwork represented a common language of working together in its own right.

Paper was, thus, everywhere. It differed from that in scientific research work, however, where documents might appear to be the goal of the work, taking the form of key commodities in what Latour and Woolgar (1979) describe as the "cycle of credit"—that is, the competition for scientific prestige. Rather, these documents and papers acted as representations of trust among human partners, especially when these partners were separated by physical location and by time. The face-to-face negotiations of meetings provided crucial moments of talk in working together, and documents were part of such practice, as they helped participants to verify the concrete and technical work achieved, and so to reaffirm their commitment to the common goal that motivated them. Documents were not themselves the *goal* of meetings (or the goal of working together), even though they might proliferate in the context of meetings.

Conclusion

Cooperation in space science missions means that people from different nation-states, different occupations, and different organizations must come together and talk about gaskets, wires, mirrors, diffraction, specifications, and so on in order to produce functioning artifacts. It is not easy, however, to engage in the practice of cooperation; the differences that accompany the division of labor bring with them potential problems—of dissolution, conflict, coercion, aging, and boredom. Working together proceeds through and in terms of these problems, but such problems are never definitively solved or resolved by participants; the social system that participants shape through the medium of the structures of cooperation never settles anywhere, even as it allows people to produce things. The practice of cooperation indeed depends on keeping these negotiations of difference going rather than on solving social problems.

In the next three chapters, I explore this practice of working together analytically. For analytic and expository ease, I separate the analysis into familiar categories of anthropological analysis: social, cultural, political, existential. But this is more a narrative strategy than a statement about practice; in practice, these aspects overlap, moving together through time and space, layers of experience and action entangled, practically speaking inseparable. I try to convey this complexity here by allowing the narrative to cumulate, common elements resonating across the chapters' linear arrangement that teases out and lays bare what in practice is intermingled.

Four

Struggling with Diversity: The Social and Cultural Dynamics of Working Together

> In Europe, we work together in a more friendly way because we have more independence. We always fear too much dictatorship; the best thing is working closely together but still being independent.
> *(A university-based PI, member of a Science Team)*

THE STRUCTURES of cooperation define the outer parameters of practice, as I have indicated, but they do not thereby render participants either silent or passive. Cooperation is, in fact, a site of struggle, a struggle for individuation as well as for connection. This struggle is brought into being by the division of labor itself. It is the division of labor that establishes the permanent, if ever-shifting, ground of working together; in so doing, it defines the social shape and cultural contours of the process itself.

The social shape of working together is one of dispersal; participants are spread out all over Europe, and across oceans and continents. People struggle against this dispersal to achieve integration not only technologically but socially as well; as participants are endeavoring to integrate the technology into a successful, functioning scientific satellite, they are simultaneously endeavoring to integrate themselves into viable social groups that can undertake the work at hand. The cultural material of the division of labor is difference, reflected in the fact that participants have distinct interests, abilities, allegiances, identities, knowledge, and goals. Such differences often engender conflict, as people emphasize what distinguishes them from each other; however, this conflict is always matched by its counterpart, the harmony that instead highlights connections among differences.

Participants in working together thus transform the static structures of the division of labor into a dynamic process. They dismantle the organization of differences, kept separate by distance and function, leaving only the differences themselves. These myriad differences appear in practice as emblems of the fundamental diversity that not only characterizes the social system of working together for participants but appears as its driving moral force. Participants strive to protect and to foster diversity as they engage in the endless dance of negotiation, conflict, and connection that constitutes working together itself.

In this chapter, I focus on the social and cultural dynamics of practices that are (dis)organized by and around diversity. I begin by looking at the significance of diversity in the social oscillation between the tendencies toward integration and dispersal, as people come together (literally and figuratively) and then pull away. I also explore the cultural values of harmony and conflict that are linked to the motivating principle of diversity. Participants assessed each of these values both positively and negatively, albeit in different contexts; conflict and harmony are in a dialectical relationship, each providing a necessary counterweight to any excesses the other may bring about. The significance of diversity that I describe here is not limited to these social and cultural aspects of working together, however. The problems and possibilities that the value of diversity generates in practice also characterize political allocations and evaluations, and existential worries and yearnings, dimensions of working together I take up in subsequent chapters.

Integration and Dispersal: The Social Dynamic of Working Together

> There are many fights, but we can always sit down and talk.
> *(A project scientist)*

> There seem to be a lot of nonintegrated modes of doing things around here.
> *(An American engineer participating on an ESA mission)*

Participants in working together perceive a kind of linear progress when they look toward the integration of the technology itself, watching it develop from ideas to sketches to engineering models to working equipment. Socially speaking, however, it is difficult to find such progress, as the people involved in mission work participate in an endlessly oscillating dynamic between integration and dispersal. The process of working together in fact depends on both centripetal and centrifugal forces: integration, the centripetal force, draws people toward one another, propelling them to share ideas and produce technology; dispersal, the centrifugal one, flings people outward, distinguishing them in work undertaken independently, separating them with distance, discipline, and office walls. Working together is defined by the interrelation of these two social forces, as it oscillates between movements toward integration and movements toward dispersal, each movement called up in response to the (near-)accomplishment of the other.

The push and pull of integrating and dispersing forces are particularly significant during mission development, since the hierarchical structures of ESA's bureaucracy do not, as I have indicated, encompass all of the participants in a neat and orderly framework. Participants are not already conveniently collected in one place, or even in one organization; indeed, the participants themselves considered the "ground state" of working together to be one of dispersal, which the social shape of the division of labor in this context necessitates. The problem for participants then becomes how to take the chaos of diversity and dispersal and construct enough order that things get done, but not so much order that the independence of participants is unduly threatened. This is a social and cultural task that depends on tension and contradiction, since the establishment of order is as much a danger to, as a prerequisite for, action.

The recognition of dispersal as the "ground state" manifested itself in an awareness, especially on the part of ESA scientists and engineers, of the importance of bringing people together, as they surveyed a field of participants who did not perceive their connection to each other. Through their roles as liaisons between ESA and various constituencies (e.g., scientists in universities and research institutes, engineers in national industrial firms), they attempted to forge links among people who pursued their mission tasks in isolation from one another; said one, "Part of our work is to be out and about." Taking the many pieces leading independent lives in a dispersed field, they "glued them together."

The term *glue* had been self-consciously adopted, and it recurred in multiple contexts. SSD scientists in particular talked about their role as one of "gluing": "We have to pull the scientists in the community together; SSD provides the glue" (I discuss the implications of this gluing role for SSD scientists' social identities in Zabusky 1992 and 1993c). Others had begun to use the term as well; for instance, an academic scientist said to me that the work of his Science Team was to "glue together a community." However, "too much glue, and people will resist," observed another SSD scientist; here was an acknowledgment that dispersal is as much a force as a ground state, a reactive force that starts up in contexts of integration. One reason integration is so important in the working-together process is that it is particularly in moments of social integration, however temporarily achieved, that technical progress is made. Indeed, people suggested that working together depended on the formation of a variety of bounded social groups, which not only provided tangible and intangible resources but were also critical in the production of technical outcomes.

On paper, there seemed to be a variety of stable, permanent groups that were in fact involved in the ongoing task of developing space science missions. Even a cursory examination of ESA mission documents, both technical and general, revealed the presence of an endlessly proliferating conglomeration of different social groups organized around specific technical or scientific questions or problems, all representing independent points of view: the Project Team, the ESA Science Team, the Operations Team, the Ground Segment Working Group, the Instrument Consortium, the Instrument Science Team, the Advisory Committee. I described such groups rather unproblematically in chapter 3, and in fact, whenever I asked people about the groups involved in mission work, they readily pulled out "organigrammes" (organizational charts) that detailed the composition and membership of ESA teams, or they showed me the tables that listed participants on different teams or consortia.

This is the structure that constrains practices, but such artifacts do not represent the practices themselves. It was in part my asking about such groups that called them into being; this is why participants could respond so readily and offer me proof of extant groups, much as they offered me proof of cooperation. In the heart of practice, however, the social groups involved in mission work were ephemeral; they came into view and then disappeared again. "Community," said one SSD scientist "is only the guys you know, sitting around a table, working together"; once "the guys" have gone off to their own offices, departments, or institutes, he suggested, whatever social group had appeared in that social space no longer existed. While engaged in the activities that constitute working together, then, people did not have an experience of bounded communities. Instead, participants' overwhelming experience was of the diverse allegiances of all players, of the openness and noisiness of diversity. Thus, contrary to the objectified "communities" of established cooperation, which were already accomplished, social groups in working together were never finally accomplished, but always immanent. People had therefore to keep making the effort to construct social cohesion in face-to-face settings, to "glue" themselves together; participants firmly believed that if people failed to develop social cohesion—that is, if they did not share information, interests, and goals—then they could not make high-quality, well-functioning technology that would answer scientific questions.

This coming together was sometimes viewed as a natural process, especially if people were left to their own devices to "figure it out," as one project scientist described it: "At the beginning, everyone was going his own way, but at the end everything got together; this process of converging happened naturally—there was no one giving orders, and that was nice." As a "natural" process, it was sometimes even likened to those processes occurring in the universe. Another project scientist described to me how he was helping the Science Team for which he was responsible come together, by circulating documents, talking to many people, and coordinating a workshop for the sharing of ideas: "This way word will slowly get out, people will start to work on things without feeling pushed or required, and then they'll come together, just like a gravitational instability at the formation of the solar system. The Community has to figure it out; ESA can't force anyone."[1]

That it is a natural process, however, does not mean it is also an easy one; after all, to continue the analogy with astronomical phenomena, space scientists tell us that galaxies, stars, and planets have "violent" births and equally "violent" deaths. Socially speaking, then, it is no surprise to find that it is a continual struggle to integrate in the face of dispersal, and vice versa. People engaged in this struggle in a variety of ways; key to this social process, as it was to the technical process, was talk in its many guises; as I described in chapter 3, talk included phone calls, electronic mail, faxes, the regular dissemination of reports and memoranda, and meetings.

Participants were conscious of the significant role played by conversation and discussion in their efforts to find connection. According to one scientist, "because the astrophysics community is all over the place, you have to keep talking"; an ESTEC engineer explained to me that "you have to discuss and exchange information all the time" in order to work well together. In fact, given the geographical dispersion and the divergence of activities and interests, "the only thing to do is to keep talking"—that is the essence of the struggle to find connection in working together.

It was not always easy to keep people talking; it took energy, commitment, and constant vigilance. I sat in on one meeting of a group of SSD scientists where the problem and potential of talking as a source of connection and involvement was debated. The issue that focused the discussion was the design and implementation of a new computer system being developed by engineers and computer specialists at another ESA site. One of the SSD scientists—I'll call him Nigel—was responsible for a data set that was tentatively to serve as a model project for this new system, yet he had not spoken to Juan, one of the developers of the computer system, since a meeting held in SSD several months before. This lack of interaction became the focus of a lively interchange, which I reproduce here.[2]

MICHAEL [the boss]: Let's talk about this new computer system and how it relates to the problem we've just been discussing. [He looked at Nigel, who shrugged.] I'm looking at the guy who's supposed to be the system's pilot project.

NIGEL: I haven't heard anything from those guys since the meeting in December [three months earlier].

PETER: [Joking.] Let's compose an e-mail to Juan right now.

MICHAEL: Look, we have to keep a bit of pressure on them, and take a more active than passive role, if we want to be involved in developing a good system which will serve our needs.

NIGEL: [Petulant.] I don't have the time. I think they should approach me.

NICO: It's a chicken-and-the-egg situation.

MICHAEL: [Direct.] Right. By having a dialogue you get the thing going the way you want, rather than finding out six months later that it's all been set and you've been left out.

NIGEL: [Complaining.] Why haven't I been involved yet?

MARC: Well, I have been involved with them through another project I'm part of; our group is talking with them quite frequently, and it's not one-sided, either—we both pick up the phone. If you don't start that, it won't come to life.

NIGEL: [Annoyed.] But I have no reason to pick up the phone because I don't hear anything.

NICO: [Exasperated.] But you're taking the salmonella approach.[3] I have no reason to talk to them either, but I know the system is there, and the time to merge things is now, so I call.

MICHAEL: Right—we have to contact them.

NIGEL: [Sullen.] Okay, I'll do it, but I don't see why I should contact them.

NICO: [Joking.] This is not the chicken and the egg, but Mohammed and the mountain.

Nigel was happy to continue working away on his project independent of other sources of advice or ideas; he enjoyed his independence, which also enabled him to retain a certain distinctiveness. His resistance to talking was an effect of diversity. His colleagues, however, were urging him to participate, to merge with the analogous activities being carried out in isolation by others; they demonstrated to him by their own example the importance of talking, all the while recognizing how difficult it was to set in motion the forces of integration. This kind of discussion was replayed in a wide variety of circumstances—in meetings, in private conversations, in memoranda that were worded to promote both dispersal and integration.

The differentiation of participants is matched by linguistic differences, as I mentioned in chapter 3, making talk as a means of integration complex. Even though participants may use one language when working together, they do not thereby cease to think in or make use of their native languages. In fact, as I suggested, the availability of multiple languages may be in itself advantageous in the process of working together. Here, I want to explore another reason why this might be so. The benefit of this particular kind of diversity derives from its role in actualizing the forces of both integration and dispersal; using a single language serves to pull people together, and simultaneously to preserve individual interests, as the following example illustrates.

During a lunch break of a mission working group meeting, I sat with a British, a German, and two Dutch engineers (only one of these people was an ESA staff member). We talked about language problems, and I asked them whether the common use of English posed difficulties or advantages for the participants. The British engineer commented that in his experience, "working in English isn't ever really a problem, because everyone speaks so well," and also because he, as a native speaker, had become accustomed to speaking "at an appropriate level." The German engineer interjected that for the most part he did not find it difficult to speak in English, although sometimes "I have to think very hard about how to phrase things to make sure I'm saying them politely."

The two Dutch engineers, whose English was virtually flawless, agreed that most people had a good command of English. Still, they expressed the view that the British did have a certain advantage, especially in meetings "when it comes to fighting, because they know how to do it." The British, these Dutchmen indicated, were able to manage the subtleties of the language, and to manipulate the rhetoric in order to persuade people of their points of view. On the other hand, they jokingly added, "this doesn't really help the British to win their arguments, because the fine points are lost on us, because we can't follow them." The British engineer agreed with the two Dutchmen, noting that "the others can get theirs back at the end if they haven't been able to fight well enough for their position during a meeting." I asked how this could be so. He explained:[4]

> After the meeting, in the corridor, they say, "Well, we didn't really understand all of that so we'll write it down for clarification." They go home and write up a report about what happened in the meeting. But when they send

> their written version around, it turns out that they have written down something completely different, something that reflects their own interests and point of view. This means that negotiations have to be opened up again!

The common use of English facilitated integration, as everyone was able to communicate at a reasonable level, worrying only about "politeness" and "appropriateness," yet this did not entirely counteract the dispersing tendencies inherent in the participants' multilingualism. As the two Dutch engineers pointed out, people could use the fact that they were speaking a language not their own to undo the integration achieved in a meeting, pulling back on the basis of an asserted need to review material in their own language. This individuating claim was itself reintegrated into the flow of working together, however, because the results of the deliberations at home in, in this case, Dutch, had again to be translated into English. Tendencies toward integration and dispersal are, in this way, both part of speaking English together.

Talking is important even for those who are not separated by geographical distance. This was the case, for example, for ESTEC engineers working on Project Teams; although members of these teams were located along a common corridor, sometimes even sharing offices, their individual tasks divided them. The social integration that created a sense of "teamness" depended on constant talk and attention to interaction. One engineer, describing the way in which, on his previous job in industry, he had worked almost exclusively by himself, talked about the constant informal interactions that characterized his work on the ESTEC Project Team. He mused: "If I were given a chance to work at home, three days a week, with a modem connection, I guess it would be more efficient, and I would get more work done. But I can see that with the type of big-scale project which we are working on here, it is essential for me to have constant access to other engineers, and for them to have access to me."

In "teamwork," as ESTEC engineers described it, "people depend on each other," and in working on space science missions, team members "all work together as equals towards a common goal." The hallmark of teamwork was, most importantly, "dropping in," the frequent, daily contact carried out in an informal way, such that "you can step into anybody's office at any time and people really try to answer your questions and don't just try to get you out the door." Rather than working "in isolation," explained one engineer, "we brainstorm together." For engineers, then, it was the ability just to "drop in" on a fellow team member that promoted social cohesion; dropping in was a force for integration.

"Dropping in" was, however, difficult to achieve for those participants whose "team" colleagues (as in the case of Science Teams, or Working Groups) were located in different sites around the world. For these radically dispersed participants, "dropping in" had to be organized more formally. It had to be organized because it was particularly in face-to-face interactions that the integrative force could be effectively realized. The frequent convocation of team and group meetings, design reviews, and other scientific and technical conferences offered one solution to this problem, and it is to such meetings that I now turn.

Meetings

> We're doing too many meetings, but we can't do with less.
>
> *(An ESTEC engineer)*

> The concept of having people going to the presentations [at the design review] is good because it keeps the community involved, and it gives the opportunity for a lot of discussion with colleagues regarding the merits and demerits of certain aspects.
>
> *(An SSD scientist)*

Nearly everyone involved in mission work went to meetings all the time—meetings were a way of life, and indeed many participants expressed their belief that in some ways meetings seemed to define the work that they do. They resented this, however, because from their perspective, meetings were a waste of time. Meetings were an interruption to work, and they interfered with work: "Sitting in a meeting is not working," one SSD scientist complained to me, yet "without meetings, nothing would get done." This scientist meant that no technical tasks would be accomplished without meetings, since people tended to pace their individual work to meet those deadlines which the convening of meetings imposed.

I want to suggest, however, that technical tasks did not provide the sole motivation for having multiple meetings; important social and cultural tasks are also accomplished in and through such collective gatherings (Brenneis 1994, Schwartzman 1989). Schwartzman (1989: 7), writing about such social and cultural dimensions of meetings, argues that meetings are "ordinary behavior with extraordinary significance." In particular, "meetings provide the organization with a form for making itself visible and apparent to its members, whereas they also provide individuals with a place for making sense of what it is that they are doing and saying . . . and what their relationships are to each other" (ibid.: 9). This can be said as well about meetings convened in the context of ESA space science missions; in this ethnographic context, however, what needs to be made visible is less the organization of ESA than the bounded shape of discrete, task-focused social groups such as teams and working groups. The participants in ESA mission work were themselves aware of this, as the following example shows. At a scientific meeting of a consortium working on a payload instrument for an ESA mission, the PI rose to make some brief remarks at the dinner that closed the event. He called attention to the presence of one of the consortium "overseers": "This guy has been shelling out money for a while, but he didn't really know if a consortium existed. So I invited him along so he could see for himself." The PI in this way highlighted the fact that only in a meeting could the social group which was the "consortium" be seen and experienced.

In the ESA context, meetings had a distinctive shape and energy, the particulars of which I describe here before considering the social and cultural dynamics of such meetings. Meetings took place in conference rooms, and the shift to meeting time and space happened gradually, as participants arrived at the meeting room one by one, or in small groups. People chatted about the weather, travels, recent conferences they had attended, people they had seen, and so on. Most people took their place around the central meeting table; more marginal participants and observers typically sat at the room's margins, along the back or side wall, signaling that they were not real members of the group and so not real attendees of the meetings.[5] When the chairperson opened the meeting—often several minutes late—the desultory conversation would die away, as the focus turned forward, toward the speaker and the projection screen at the front of the room.

Information in meetings was conveyed not only by speakers and discussants, but by both handouts and transparencies. Presentations were balanced with open discussion; the chair, as facilitator, did not, and could not, impose any kind of formal structure on talk, except to solicit contributions from those who perhaps had not yet had a chance to speak, to put time limits on discussion, since the clock was always ticking, and to provide summarizing statements between agenda items. Generally, particularly in meetings of working groups, discussion was lively and involved nearly everyone, sometimes in a disorderly tumult, other times in more orchestrated exposition. During discussions of specific technical areas to which not everyone was capable of contributing, a few members of the group might dominate the discussion; as one engineer told me, "Meetings have parts I don't understand or I don't care about." Reflecting this diversity of interests, sometimes meeting participants would silently rise and leave the room for brief periods while the focus was on something not germane to their area of expertise or their interest.

Meetings could last a few hours or several days. Typically, daylong meetings were broken up by coffee breaks in the morning and afternoon (although coffee was often drunk throughout the course of the meeting), and lunch breaks at midday. However informal a meeting might be, these breaks offered participants a respite from its pressure. At lunch, often members of the group would sit all together but turn their attention away from technical details and difficult issues raised in the meeting, discussing general policy, other scientific conferences attended recently, people, travel, national customs, language, and so on. Sometimes, during a multiday meeting, groups whose members came from diverse locations would also arrange to eat dinner together. All of these leisure-oriented gatherings in the midst of a meeting that spanned two or more days were essential parts of the meeting itself.

Meetings often closed with heated discussion, as participants tried to decide when and where the next meeting or meetings should take place. The chair typically proposed one or two possibilities, which were invariably challenged by one or more members of the group. All participants then put their available dates into the central pot, and intense discussion ensued as to which date was the best. These discussions could last as long as half an hour; people grew adamant, passions would rise, and all pretense at collectivity often disappeared. One scientist likened these interactions to those taking place on the floor of the stock exchange. After such debate, meetings often ended as informally as they had begun, with people departing one by one to catch planes and trains, as the focused discussion around the table dissolved into more intimate conversations.

As I have said, the ongoing process of mission development generally occurs in a context of dispersal, which is entailed by the differentiation of participants. Meetings punctuate this isolating dimension of the process, providing one of the primary opportunities for people to come together, to "drop in" on one another, to trade information in a face-to-face forum. Meetings provide the occasions for, as Durkheim would have it, "the spark of a social being" to be ignited; in meetings, participants experienced social integration and the affirmation of "teamness" or "groupness." Thus, meetings, although viewed as wastes of time, cannot be dispensed with. Such gatherings represent the forces of integration at work, for it is during such face-to-face encounters that the collective dimension of working together is made manifest and directly experienced.

Meetings are also important for those who *do* work in the same place, whether or not on the same missions, since even here, differences (e.g., of discipline) can effectively isolate members of a single department. For this reason, project engineers attended regular, weekly Project Team meetings, the purpose of which was expressly to communicate, rather than, for instance, to make decisions. This was also the case in SSD where the staff scientists' work separated them from each other; it was only in the regularly scheduled divisional meetings that any kind of cohesiveness could be generated. One division head explained that he held meetings for all the staff in his division for just this reason: "I am trying to bring people together so they can realize they're all one entity, working together."

In this section, however, I am talking primarily about meetings that are part of mission work and bring together participants from diverse locations and disciplines. The case of the design review provides an especially good example of this kind of meeting, meetings that are, as I have argued, not simply "technical," but culturally and socially significant as well. Design reviews provide opportunities for the evaluation and assessment of ongoing development work; as such, they are some of the key "milestones" in mission schedules.

As I have indicated, the division of labor accounts for the ground state of dispersal in working together. As part of the technical division of labor, participants break down satellite technology and systems into distinct parts; people work in a very focused way on those parts which are their responsibility. This breaking down means that, for instance, an engineer in industry working hard on the thermal design and manufacture of a particular subsystem has no occasion even to think about the other subsystems being developed, let alone to talk with those working on other systems; he or she will certainly not be concerned about the payload, even though it is the payload that will be carrying out the scientific goals of the mission. In part, this is an effect of the specialization that accompanies the division of labor.

In the context of mission development, a consequence of this fragmentation is that a kind of silence envelopes the whole. The technological "whole" is seldom seen; indeed, for long periods of time, it does not exist anywhere except in designs, which are themselves broken down into myriad components. Just as this technological and scientific whole is absent, so too is a social "whole." There was, in fact, no native term to describe the social collectivity that is involved in mission work. *Project Team*, for instance, designated only one small group of people;

there was no way to talk about everyone, from ESA engineers and scientists to external ones, from ESTEC staff to ESOC staff, from payload designers to spacecraft developers. The term *mission* referred to the thing itself, that which was being designed and constructed, as well as the scientific goals and scientific data to be collected. But the "mission" did not have any people in it. There was no such thing, in this world, as a mission "community"—it was too amorphous even to warrant a name. People focused on discrete aspects as they built integral but isolated components—and there was no one person who saw it all, or who could control it all.

Design reviews intruded into this steady flow of the "exigencies of life" (to borrow from Durkheim) with intimations of that silent, missing "whole." These reviews provided a momentary breathing space in the headlong rush to deadlines; they were interruptions, creating a social hiatus when time was suspended and the gestalt of the mission was encouraged to emerge. During the meetings that accompanied these reviews, the diverse participants, hard at work on the mission, were momentarily integrated as they, for a period of time, sat next to one another, talked to one another, and ate with one another. In coming together in this way, they established their cohesiveness and forged connections in terms of the technology itself.

In design reviews, held often at ESTEC, representatives from industry and other laboratories joined with ESA participants to review the status of spacecraft and payload development. These reviews included many hours, over one or more days, of public presentations and discussion, both of which were carried out in great technical detail as people went over the "nuts and bolts" of the technology and satellite systems. Presenters and auditors at such reviews included industry engineers, members of the Project Team, engineers from other ESA divisions and departments (including the Technical Directorate), and members of the Science Team (including the project scientist, mission scientists, and PIs). Other interested participants might attend, depending on the particular systems being reviewed; for instance, at a review of the ground system design, those people working as part of a Ground System Working Group might attend.

The gathering itself, however, was not a forum for evaluation. Members of the evaluative board were expected to review carefully the thousands of pages of documentation that accompanied the presentations. They were to make their recommendations based on this written material and *not* on the presentations, which provided only distillations of the details. Yet the public presentations were at the same time considered to be an essential, indispensable part of design reviews. This was a paradox, and it pointed to the fact that participants were tacitly aware that such collective gatherings were about more than technique and technology.

Participants told me that the design reviews in fact had a necessary "psychological function": they provided participants with what they called "perspective," and they helped to generate "team spirit." According to participants, these meetings accomplished this by reminding everyone of the "big picture," and by demonstrating how all the technical parts were interconnected such that decisions made in one small section might have significant repercussions elsewhere in the satellite. In this way, these gatherings made tangible a social group of people working on a common set of problems. They showed people that their work was connected to the work of myriad others, that although they might be laboring in isolation, their work was not itself isolated. The sense of connection that people derived from these encounters could be carried away, as they dispersed to the local and separate environments of departments in industries or institutes.

The emphasis on integration was made clear in the opening presentations of design reviews. For an hour or so at the start, a few speakers who were affiliated with the mission (rather than with a given instrument or system) provided information on the "mission overview," giving people a sense of the ultimate goal toward which they were working. Such an overview included presentations by ESA project engineers concerning the planned shape, size, and constellation of the complete and integrated satellite; in particular, the engineers talked about the makeup of the payload, where and how the payload instruments fit into the spacecraft, how the ground system communicated operations data to the satellite and retrieved scientific data from the satellite, and so on. The overview also included a presentation by the project scientist on the mission's scientific goals (its raison d'être), which would be accomplished by the complete and integrated satellite. Design reviews, thus, began with a vision of completed and integrated technology that corresponded to the social integration of the human participants.

But integration did not mean the elimination of diversity or the end of dispersal. Inside such meetings as design reviews, the centrifugal force of dispersal made itself felt in the majority of the presentations, which focused closely on those individual components and systems developed in isolation. In giving these presentations, speakers clearly identified themselves according to their occupational and national differences—as academic scientists or project engineers, as ESA or industry representatives, as French or German citizens. These differences were expressed by the content and style of their presentations: what details and subjects they chose to focus on, the manner in which they carried themselves, how they dressed, the quality of their transparencies, and their facility with and accent in English. People noticed these differences, and they accepted and even enjoyed them; they were not impediments to connection but one source of the possibility of connection.

Even though meetings are, in this way, integrating moments, however, they do not proceed in an atmosphere of loving-kindness, where camaraderie unproblematically supersedes hostility. Meetings encapsulate both conflict and harmony, in part because they give room to difference even while forging connections. In the next section, I turn my attention to the interplay of conflict and harmony in meetings, and their connection to the social forces of dispersal and integration.

Conflict and Harmony: Cultural Trajectories in Working Together

> I think that it's inevitable that with five different groups and five different people there will be five different ways of doing things.
> *(A PI, member of a Science Team)*

> You try to come to some kind of compromise which satisfies everyone—or dissatisfies everyone.
> *(A project scientist describing a Science Team meeting)*

Meetings, as integrating moments that contain diversity, are particularly volatile occasions in working together because in them differences come face-to-face and must simultaneously accommodate and resist each other. Such close quarters of social connection accordingly motivate the expression and actualization of both harmony and conflict. Harmony and conflict are the alternative cultural trajectories that a commitment to diversity can take. Harmony expresses the value of difference as interdependence; harmony talk is about consensus and compromise. To actualize harmony means to get along by recognizing what connects diverse interests. Conflict, on the other hand, expresses the value of difference as independence; conflict talk is about control and individuality. To actualize conflict means to get along by recognizing—and respecting—what distinguishes interests.

These cultural trajectories resonate with the social tendencies of integration and dispersal, which they resemble, in that integration and harmony both involve connection, while dispersal and conflict both involve autonomy. But these are, nonetheless, independent. That is, it is possible for people to come together, for instance, either in a spirit of harmony or through spirited conflict. Similarly, people can disperse to their individual sites either harmoniously or in conflict. People are ambivalent about these twin values as a result, unable to privilege one at the expense of the other, despite the rhetoric of cooperation that highlights harmony as a pragmatic approach (in harmonization) to achieving both connection and profit. This rhetoric, like the structures that accompany it, encompasses working together, and it is particularly present during moments of integration such as meetings, when connection resonates down from structure through to practice, and vice versa (I return to this relationship between cooperation and harmony in chapter 7). This is one reason why people felt dismayed when there was too much or too intense argument during meetings; meetings were "supposed" to provide opportunities for "converging to consensus," and not, as it were, for diverging to independence.[6]

But in the temporal unfolding of practice, no single trajectory can dominate the scene, given the complexity of differentiation and the dynamism of the process itself. And in fact, people welcomed conflict, even encouraged it during meetings, because they recognized that they could connect with each other only by combining differences, a combining which necessitated the enunciation of individual interests during meetings. Harmony and conflict, then, were both desired and both criticized simultaneously; what it comes down to is that both are necessary in the expression and protection of diversity in working together.[7]

When people were riding the cultural trajectory of harmony, they talked about "consensus": said one project scientist, "In working together, we converge towards consensus." Consensus is, indeed, a mechanism of harmony; it is what brings people together, culturally speaking, as people expressly try to convert their independent interests into a common good. For instance, sometimes during Science Team meetings, all of the participants joined with each other in expressing their commitment to the social group that had appeared around the table. In this way, they indicated to each other that they had left their independent affiliations and distinctions at the door, as it were, and had become members exclusively of the Science Team, whose shared interests brought them together in compromise and, perhaps, in opposition to those they might otherwise represent on the outside. During these discussions, the door was closed against these outsiders; this was the time for harmony talk, in which people interacted in a spirit of real give-and-take, sharing information and collectively solving problems by incorporating the diversity (of expertise, of experience) of participants. Typical discussions of this kind arose when, for example, one PI asked for ideas on how to solve a problem his consortium was having with cracked gaskets, or when a PI expressed his desire to try to test a silicon semiconductor component—one that everyone was using—together with the others, in order to save time, effort, and money. Such discussions proceeded in an atmosphere of good humor, as people gave each other mutual support, pointing out possibilities, clarifying issues, offering assistance. Even when there was questioning or criticism, it was often not expressed negatively or aggressively, but rather in a spirit of common inquiry.

When people were riding the cultural trajectory of conflict, however, they talked about "autonomy" and redefined consensus as "coercion." In response to this perception of the threat of unilateral action, participants gave voice to their independent interests. The assertion of interests is a mechanism of conflict; it is what distinguishes people, culturally speaking, as they resist their subsumption into a common good. For instance, to take the case again of Science Team meetings, sometimes the PIs emphasized their roles as representatives of instrument consortia. During these interactions, what surfaced were the independent interests of participants, interests that divided people from each other. The PIs spoke as if the other social groups of which they were nominally members were present in the room, a coterie of invisible listeners hanging on and judging every word. This was conflict talk, in which a recital of individual points of view replaced the give-and-take of truly interactive discussion. This happened particularly when the topic being addressed concerned a sticky policy issue, a philosophical disagreement, or a recommendation for action. These issues necessitated joint action or compromise, and so required that all meeting participants had the opportunity to express their points of view before they could be expected to join any agreements.[8] In enunciating these diverse and separate views, people recited toward a center rather than interacted through it, the center sometimes being physically located in the seminar table around which people were seated, other times in the person of the chair wherever he or she might be seated.

When the emphasis is on harmony, independent interests appear not as justifiable expressions of autonomy but as "self-interest," as parochial wants, as selfishness. During such moments, participants evaluate conflict negatively. Conflict appears as that which renders work inefficient and potentially unproductive; it rips apart the center and leaves disparate pieces and people at odds with one another. In harmony-centered discourse, conflict is indicative of a kind of childishness, as people allow their petty desires to interfere with the positive work of building connectedness technologically and socially. Harmony, on the other hand, appears as unequivocally positive, leading people to unity, where they can work efficiently and productively, without wasting time, energy, or resources. In the positive valuation of harmony, the fact that people can willingly give themselves up to the common good is viewed as a mark of maturity, wherein people accept a kind of delayed gratification that brings a different kind of immediate reward—namely, the attainment, however temporary, of social connection.

Invocations of harmony appeared particularly during moments of intense argument, when voices were raised and people banged on tables; in the midst of such acrimonious debates, somebody invariably called up the image of the common good in order to counteract the dissolving consequences of conflict, announcing, "We're all in the same boat" or "We're all working towards a common goal." Such interjections reminded participants that differences could be united in interdependent form, that argument was for sharing interests, not protecting them.

These moral pleas point to what people have in common rather than what divides them; as such, they are difficult to argue with, especially in the integrating moments of meetings: "It is very much in committees," said one SSD scientist, "where you are airing contentious issues, including those which you know will be anathema to other people, but you have to raise them and work them out." This is why meetings can combine humor with intense and angry debates. One SSD scientist described the meetings of the Science Team he worked with as "always fairly humorous, because although it may not be easy [to work together], we're all in the same boat." He preferred to focus on harmony and consensus; his meeting minutes almost always omitted discussion of any sharp or heated exchanges that might have occurred, even if those exchanges dominated a significant portion of a day's events. Instead, he summed up the diverse points of view represented in the argument, such that the minutes reflected diversity, not divisiveness.

This project scientist occasionally actively worried about the sharp exchanges that sometimes occurred during meetings; in one case, following one such meeting, he told me of his concern that an argument which had taken place during the day's meeting might "cause relations to deteriorate." To his relief, the opposite happened; after the argument was over, "everyone wanted to show they were being constructive and working towards a common goal," so in the end the heated exchange facilitated the movement toward consensus. This is harmony in action.

When the emphasis is on conflict, however, such consensus appears not as an expression of interdependence but as conformity, and the individual interests condemned in consensus become the necessary antidote to such pressures toward homogeneity, appearing as individual rights, as the source of diversity itself. Harmony instead is that which renders work bland and uninteresting; it suffocates possibility and creativity by eradicating diversity. In the conflict-centered discourse of participants, harmony puts diversity in chains by insuring the conformity of shared interests. Conflict, on the other hand, is evaluated positively by participants because it leads to freedom and independence, allowing people to work on the cutting edge, where the quality of ideas is improved because of the expression of different perspectives. In the positive valuation of conflict, the fact that people have and maintain individual interests bears witness to their fundamental equality.

Those who are involved in working together thus tended to perceive harmony and conflict as irreconcilable opposites. They could see what was good in both yet also resisted the implications of both, as I have suggested here. To see conflict and harmony as opposites, however, is to see them as independent of each other, but they are not, as Nader (1990: 308) also suggests, "antithetical." They are, in fact, codetermining—they *require* each other. It is my argument that the genesis of each is embedded in the accomplishment of the other. That is, it is precisely during meetings, occasions for the integration toward which everyone works—which people and technology need to make happen, and which people foster through expressions of harmony—that people must also pull away and assert independent interests, divergent philosophies, contrary opinions. In this way, I suggest that people acknowledge the need to initiate as well as to complete, to disperse as well as to integrate. But such invocations and expressions of conflict, and of the openness and equality it entails, themselves call into being harmonizing discourse, as people struggle to pull together, to find connections within diversity by combining interests, sharing philosophies, and exchanging opinions. This is a relationship less of static symmetry than of dialectical dynamism; conflict and harmony, by calling each other into being, keep everything moving in and around the diversity from which they spring.

Conclusion

In working together, people revel in diversity, and seek to preserve it, even as they struggle over and over again to find and make connections among the many disparate pieces of their practice. Meetings are a manifestation of this drive toward integration; they give people the chance to make connections, and to see connections, both in the technology and in their social relations. Social effort, during these temporary gatherings, is turned toward the deconstruction of boundaries, in order to facilitate such connection, but the deconstruction is not accompanied by eradication of differences. This is a crucial distinction, because it illuminates the way in which participants are committed to diversity in the practices of working together: differences are not annihilated but combined, and in this combining they are respected, even honored.

This combining of differences is facilitated in different ways. During harmony moments, as people try to draw a boundary around themselves in an effort to manifest a kind of "groupness," participants downplay their differences even though they may not eradicate them. For instance, they may deny that nationality has a relevance in working together, or they may emphasize the way in which they "all work as one," as the statements about working together in chapter 3 demonstrated. The emphasis on harmony entails the fostering of sharing (of information and interests, of perspectives and goals), but it does not require unanimity. This is why people try to "converge to consensus"—it enables people to make a common ground without giving up their independence and without losing themselves in solidarity.

Conflict arises as a necessary counterweight to harmony during moments of integration, because it gives expression to the independence that people need if they are to maintain their distinctiveness even when coming to agreements. For this reason, the social groups that people make in integrating moments do not represent solidary bands of like-minded members; instead, they contain the conflict of (often irreconcilable) differences. Indeed, for social groups to be viable in this milieu, they *must* include and contain distinctions within them; "community," whether scientific or European, is a structure of conflict that has been forged in the crucible of working together, a process committed to and permeated by diversity.

The awareness of social communities' conflictual and temporary quality was often expressed by scientists in SSD who, when I asked them directly about the existence of the "scientific community," insisted that there really was no such uniform social space as the "scientific community." They talked instead about divergent and conflicting interests, about scientists competing for scarce funds and opportunities, about the "cutthroat competition" that characterized relations "out there." They depicted fluid networks of relationships, rather than a single, bounded group, networks that coalesced and dissolved, expanded and contracted, depending on the situation and the need. Yet this kind of awareness was tempered in action; for SSD scientists, the scientific community was also a social and cultural fact, and they often spoke as if it were in relation to it that they pursued their individual careers and executed the tasks of mission support work. So it is that SSD scientists could simultaneously invoke an accomplished entity and deny its materiality or, at least, its normative value; the "scientific community" was not, to them, a solidary band but an arena for shared conflicts.

Coming together and yet remaining distinct, the characteristic of social practices in working together, does not appear to be an impossibility for European participants on space science missions, although it might appear to be to others, for instance, to Americans (see Greenhouse 1986). Hervé Varenne (1977: 91), discussing the idea of "community" he found expressed in a small American town, noted that people there worried about how to maintain the social order in the face of the "fluidity" of "shifting communities." One way they accomplished a sense of order was by damping down their differences, by not allowing individual interests, for example, to intrude into public domains. Indeed,

> It seemed very difficult for most people to accept the idea that they might be unique or even in a minority on a particular subject. Any position had to be legitimized by reference to the fact that "everybody" held it to be valid, not only the actual group in which the position was uttered but any intelligent, sincere person. Some of my informants could literally be brought to tears if someone they considered close to them appeared to seriously disagree with them. (Ibid.: 92)

This despair that accompanied statements of disagreement was most decidedly absent in the European contexts I am exploring here. Conflict was everywhere, lurking in corners, erupting in meetings, in corridors, in an impassioned exchange of letters. The disagreements of conflict were primarily expressions of distinctiveness, and as such they were socially necessary, culturally valued, and even exciting. This was an effect of a practical commitment to diversity, even as it might also reflect, as I argued in chapter 1, the European ideology of "unity in diversity."

Building a scientific mission depends on integration in part because it is in such collective moments that important decisions are made, or at least are made public and thereby legitimate. Establishing agreement, which the making of decisions requires, becomes a critical problem in working together, because such agreement threatens to still the ongoing oscillation between integration and dispersal that gives life to the process itself. In decisions, there is the intimation of unanimity, of the silencing of diversity and the imposition of conformity. For this reason, people struggle for control of these moments in order to preserve their independent interests—interests that include professional autonomy, the integrity of distinct components and instruments, and the rights of departments and institutions to be free of ESA's control and to determine their own actions. In working together, then, the making of decisions is a highly political moment. It is to this political dimension of practice that I turn in the following chapter.

Five

Evasion and Responsibility: The Politics of Working Together

> This mission is more complicated and more of a mess [than others I've worked on], basically because it has more participants. . . . In theory, the Project Team is supposed to be managing everything, but in practice, we all work together.
>
> *(An SSD scientist participating on an ESA mission)*

IN THE PRECEDING CHAPTER, I showed how the division of labor is experienced in the practice of cooperation—as a field of diversity out of which participants must build social cohesion even while maintaining the integrity of differences. The integrating moments of working together are bound up in particular with the making of decisions, because each depends on establishing connections. But to establish such connections, it is not enough simply to be together, hoping that out of simple contact things will emerge. The making of things requires the exercise of both power and control; these are, fundamentally, political issues.

When I say "politics" here, I am not concerned simply with the allocation of material resources; politics includes also, as Myers and Brenneis (1984: 4) argue, "those processes involved with the regulation and allocation of social value." Myers and Brenneis argue, indeed, that "a polity is as much the creation of meaning as of sanctions or coercion" (ibid.: 11); this turns the analyst's attention to the significance of talk, since it is those who "get to be heard" rather than those who "coerce" (ibid.: 12) who can effectively influence decision making on both material and symbolic matters. From this perspective, political struggles occur as much through noisy negotiations of meaning and legitimacy as they do through material and physical coercion. Indeed, in working together, political struggles are undertaken primarily through the discursive and evaluative categorization of people and activities, as participants strive to articulate positions and commitments in such a way that they can simultaneously wield influence and avoid the influence of others.

Decisions are the quintessentially political moments in working together precisely because they provide the occasion not only for the exercise of power, but also for the exercise of control. In decisions, participants must commit to particular courses of action in order to accomplish the consensual goal of getting things done; the question that is noisily contested is who has the right to mandate such commitment and therefore to direct the action of others, whether by assigning tasks, allocating resources, or determining technical requirements. People struggle both to claim this right and to evade others' assertions of legitimacy. For participants, this struggle centers on the problem of hierarchy.

On the one hand, participants accept that hierarchy is a part of working together, at least in its practical-technical form of increasingly complex and integrated tasks and components; in this hierarchy, the exercise of authority to insure integration is recognized as necessary, since without it, nothing would ever get done. From this perspective, indeed, equality has certain limitations as a principle of organization permitting collective action (even though it is valorized as a moral principle), since it seems to make legitimate authority untenable, thus giving too much room to dis-organization or dis-integration. On the other hand, the emphasis on hierarchy in technical integration carries with it the threat of domination, in part because the authority exercised in moments of decision seems, in the view of participants, temporarily to suppress diversity in a social hierarchy that enforces the conformity of obedience. Participants cannot allow the resulting sameness, and the hierarchy of control that produces it, to prevail; to them, the eradication of differences among equals threatens the process of working together entirely.

One consequence of these contending views of hierarchy and authority is that it is no easy task actually to exercise control, or even to determine who has the right to do so in working together. These are intensely political issues, in which, as in the social and cultural dimensions of working together, diversity plays a key role. Indeed, it is the valorization of diversity that makes the continual intrusions of social hierarchy problematic for participants, who wish instead to maintain egalitarian relations. Such systems, as Flanagan and Rayner (1988: 2) argue, require constant "participation and vigilance" on the part of their members, because egalitarian relations are not given, "but are achieved and maintained by the social and symbolic manipulation of often complex rules systems governing decision making." In working together, these "rules systems" are discursive systems that provide participants with the means to negotiate the politics of control (i.e., who can have it, when they can have it, over whom they can have it).

In this chapter, I explore the political dimension of working together, paying particular attention to participants' discursive construction of a social system that valorizes equality by continually undermining social hierarchy and resisting assertions of authority. In the first section, I discuss the way in which people negotiate their positions on the uncertain ground of ever-shifting social hierarchies. This negotiation takes place through a discursive system of evasion, which enables participants both to assert superiority and to escape from others' assertions of their inferiority. It provides, thus, a cultural mechanism for the continual undermining of value hierarchies that the valorization of diversity demands. In the second section, I show how people resist the establishment of a permanent hierarchy of control by keeping authority circulating. They do this both by asserting the value and importance of "responsibility" as a kind of passive authority, and by resisting the imposition of authority when it does occur.

These discursive mechanisms reveal the entanglement of the politics of working together with the social, cultural, and existential dimensions of practice; the political is not more fundamental than the social, nor is the reverse true. When people think about hierarchy, they are also thinking about integration; when they worry about control, they are also worrying about death. This entanglement can be seen clearly in decisions. These are moments that seem to push participants, in linear fashion, toward some culminating resolution, yet they contain, as well, a dis-integrating movement. The moment decisions are made, people begin to unravel their consequences and scramble to undo the status hierarchies and to undermine the control these manifest. In this way, although decisions may produce technical and social structures, they simultaneously reveal the fundamental contingency of the social order. In this way, too, although decisions seem to threaten the liveliness of the social system, they also regenerate it.

The Discursive System of Evasion

[During an interview with an SSD scientist.]

ME: What is the difference between scientists and engineers?

SCIENTIST: Well, the story goes that scientists should have the big picture, understand why something was being done and be able to interpret the results. The engineer, on the other hand, knows how the hardware fits together, can design the setup, but won't know what it's all for. But I think it's arrogant to make such distinctions. In fact, I prefer working with the engineers because they have their heads screwed on right and really know what they're up to.

[In the canteen, in conversation with a different SSD scientist.]

ME: What is it like working with the Project Team?

SCIENTIST: Look, engineers are intellectual dwarves, but I guess they're all right to work with. [He winked.] I said this for your benefit.

Are engineers "intellectual dwarves" or those who "really know what they're up to"? Are scientists intellectually superior or simply "arrogant" and perhaps deluded? Such contradictions in definitions and ascriptions emerge in working together, because the division of labor means also that diverse value systems are put into play at every moment. Everyone who participates on ESA science missions is also involved in myriad other projects and is concurrently a member of other distinct groups and institutions, most of which are entirely independent of ESA and the missions it coordinates. This is why people talked about "wearing many hats" when they described their activities. Nearly all participants had multiple allegiances and multiple memberships, each of which had its particular constellation of interests and ideas. Most people were not uncomfortable with this complexity in their social identities and cultural orientations; they believed that it added significantly to their ability to be "flexible," which was also perceived as an asset in technical work.

The division of labor means, then, that people bring to working together different cultural perspectives generated in other social fields. These fields, while distinct, are not mutually exclusive; they are, to use Moore's (1978) term, "semi-autonomous," which means that there is overlap in terms and values among all of these social fields, an overlap that generates ambiguous meaning in the social field of working together.[1] The hierarchies of cultural values generated in these diverse social fields, which render the immediate environment orderly and comprehensible, are not directly meaningful outside their proper domains. The participants in working together thus find themselves constantly contesting and negotiating the meanings of similar sets of terms and values. This is not simply a negotiation for understanding, however; it is fundamentally a negotiation for control. All participants attempt to impose their own values on others even as they simultaneously evade such imposition from others.

The ambiguity of meaning contains particularly an ambiguity about superiority—participants are constantly debating (indirectly and directly) the definition of who and what is superior and inferior. In such debates, everyone is endeavoring to escape from inferior statuses. Although the ambiguity entailed by this constant evading of other peoples' definitions is both confusing and frustrating for participants, I would argue that it is also liberating, because it prevents any single value hierarchy from becoming firmly established and so maintains the fluidity of working together. In a unitary status system, some people would be definitively designated as inferior; this would be an intolerable state of affairs, however, because working together valorizes equality in diversity.

Ethnographically, the system of evasion manifests itself in a multitude of conflicting and contradictory statements about who people are and what they do, statements that are not simply descriptive but judgmental. Sometimes these statements are made by different people confronting each other in the middle of an argument; sometimes they are made by the same person, but in different conversations; and at other times, they are uttered by the same person in the same conversation. It is not a question of inconsistency, but of context shifting. Indeed, the contradictions are best understood as different forms of evasion; as such, the contradictions reflect the effort people make to undermine the value hierarchies asserted by others.

In particular, the system of evasion is constituted by three discursive mechanisms: deflection, inversion, and reorientation. These mechanisms, each of which is expressed through varying kinds of value oppositions, operate in a social and cultural context in which people are attempting to regulate and allocate social value. These are, then, fundamentally political maneuvers, in which the struggle to define a hierarchy of values is a struggle to determine who has the right to claim legitimate authority, and so to exercise control over others.

Deflection

In deflection, people evade others' assessments of relative social value by challenging the relevance of the social field in question. They say, in effect, "Well, what you say may be true, but this situation is not about X, it's about Y, and in Y, this opposition is not meaningful." Such statements do not have to be made in a single moment of conversation, with different people taking different positions. The contexts that affect particular moments of talk can be implicit, as will become clear in the following example of deflection.

In order to make this example intelligible, I must explicate the context of the contradictory statements at issue. By way of prelude to this example, then, I offer a brief summary of one of the prevailing views that scientists in ESA tended to hold. At ESTEC, where technical tasks dominated people's attention and energy, scientists had many disparaging things to say about the work of management, especially as it was executed by those they considered to be "administrators." They expressed contempt for anything emanating from "HQ" in Paris; there, the concern was with "politics," which most SSD scientists perceived as, at best, irrelevant to their own technical work or, at worst, inimical to it—as when "political interests" determined what kind of mission should be done, or who should work on it. Thus, this opposition was not made in a balanced, symmetrical fashion (i.e., "We do science and engineering, they do management"); to talk about "science" and "management" was to make a judgment and not simply a comparison. For these scientists, as for many ESTEC engineers, the social field within this opposition that was relevant was one oriented toward the production of high-quality space technology; thus, they perceived the "technical" issues to be fundamental to the process in which they were engaged.[2]

This valorization of professional activities, one that was familiar to everyone in ESA, provided part of the contextual backdrop for an interview I conducted with a senior ESA administrator based at HQ, herself a lawyer by training. I began the interview by describing my research project (the inevitable "European cooperation in space science" of my many introductions), thereby implicitly associating myself with the scientific perspective prevalent in SSD. During our conversation, I asked for her opinion about why ESA seemed to be so successful in achieving international cooperation.[3] She offered this explanation: "It's the nature of what we're doing; we recruit engineers, scientists, lawyers, and not civil servants." Here was the critical opposition: lawyers and technologists were alike *and superior* to "civil servants" (and also, by implication, to politicians). The superiority of these actors stemmed from their commitment to the execution of technical tasks; it was from the focus on the technical—whether, the suggestion was, in technology, science, or the law—that the ability to cooperate successfully derived. In this respect, civil servants were inferior because their commitment to national rules and regulations interfered with this kind of streamlined focus. This opposition was meaningful only in a context in which the social field at issue was geared to the production of high technology for space missions.

As if aware, however, that her own participation in this social field rendered her vulnerable to the disparaging assessments of scientists, moments later this administrator changed the context. She deflected the (implicit) context of her statements from one in which high technology was the goal to one in which political cooperation was the end sought by participants. She did this by comparing the work of scientists and engineers at ESTEC to that of the administrators at Headquarters; she concluded, "It's just more exciting at HQ. The people at the technical end *don't* see how decisions are made, and just don't understand the complexities involved with running ESA." Of course, "when you're dealing with technical people you can't expect them to express themselves as if they have a law degree from Cambridge," but still, she added, "everybody at HQ thinks of ESTEC as a bunch of farmers out there."

Here the critical opposition changed; it was now between lawyers, who were superior, and "technical people," both scientists and engineers, who were inferior. Civil servants were not even relevant in this evaluation. The use of the term *farmers* suggested the reason for the technologists' inferiority; because they were out working in their fields, as it were, laboring to get technology to grow, they were unable to pay attention to, or to understand, anything else. As a result, they lacked the sophistication and political savvy that were necessary to make ESA successful. This opposition was meaningful only in a context in which the social field at issue was geared to the production of political cooperation among European nations.[4]

The two social fields implicit in this administrator's evaluative statements are distinct; although there is certainly overlap, the participants differ. That is, in one social field, the important distinction was between lawyers and civil servants, while in the other it was between lawyers and technical people. In this conversation, the administrator began by asserting one social field as preeminent but transferred the focus to another social field in order to demonstrate unequivocally who was "really" on top of the social hierarchy. Thus, what is significant is not that she made contradictory statements within a single context (i.e., in the social field of ESA, lawyers are both superior and inferior to technologists); rather, she shifted social systems altogether. She said, in effect, "It's true that scientists and engineers know more about the development of successful technology, but that's not the social system that is important; the one that is important is the one where we are developing successful cooperation." In this deflection of (implicit) social field, the administrator was able to evade the impositions of others' viewpoints, and so to undercut any negative judgment about her own work.

Inversion

In inversion, on the other hand, people agree on the social field in which they are working, but disagree on the assessment made of different tasks, terms, and values. In these discursive moments, people agree about what the significant oppositions are (e.g., scientists versus engineers), but not about their relative value. That is, on one level, and in some statements, these oppositions may appear to be balanced and even symmetrical (e.g., some people are scientists; some people are engineers). But on another level, one of the distinctions appears as superior, itself encompassing (Dumont 1986) the opposition (e.g., engineers can do their work only in a scientific context that assigns them particular tasks).[5] What is at issue in inversion is the direction of the value hierarchy; while some may say, for instance, that science is superior to—that is, "better than"—engineering, others contest this imposition of inferiority and invert the direction of the hierarchy, insisting that engineering is superior to science. Although no one disagrees about the terms of the debate, they do disagree about the value of these terms.

I have used the abstract example of scientists versus engineers in the discussion above because that is, indeed, one of the most significant categorical oppositions in working together, as I discussed in chapter 3. As I indicated there, when participants in working together talked about the differences between scientists and engineers, they were talking less about particular individuals than about an opposition between two fundamental values that guided their social process: dreaming versus pragmatism.

Pragmatism, as I showed in chapter 2, characterizes the ESA ideology of cooperation; it also appears as a significant value in the local domain of working together. I use *pragmatism*, which is not a local term, to refer to the way in which people pay attention to the myriad details that are not explicitly scientific or technical; such details include those stemming from bureaucratic regulations, economic considerations, political maneuvering, and industrial relations. That is, "pragmatism" refers to the principle of taking constraints into account in the execution of technical tasks. On the other hand, "dreaming"—this *is* a local term—refers to the principle of focusing on the technical details at issue and pushing them to their limits without regard to worldly constraints. In the context of working on ESA space science missions, scientists and engineers negotiate the relative value of dreaming and pragmatism. In this environment, it matters a great deal whether it is better to be a dreamer or a pragmatist, particularly because the ideology of ESA is one that rates "pragmatism" more highly.

At ESTEC, the standard cultural definition of scientist was as "dreamer," while the engineer was the "pragmatist." Scientists often argued that it was better to be a dreamer and so assessed dreaming as the superior value, despite the pressures imposed by the rhetoric of cooperation. In so doing, they were asserting a value hierarchy that allocated to them the superior position. They insisted that it was better to be able to focus on technical details and not let the "practicalities of life" (which they thought of as the "trivialities of life") get in the way: "Scientists have to keep out of the nuts and bolts because we have to do physics and can't clutter up our minds with other details," said one

SSD scientist. ESTEC engineers, too, at times expressed this view, even though they thereby risked being rendered inferior, since they were more closely identified with ESA and its associated rhetoric of cooperation: "Scientists have brilliant dreams and we need those people who have scientific interests and who are pushing knowledge for mankind." Many of the scientists reveled in their status as the "visionaries" of the working-together process; one university scientist, a member of a Science Team, expressed his conviction that "the attitude of dreaming is the most important one there is"—in it was "poetry," without which, in essence, life would not be worth living. It was when dreaming was in this way assessed as the superior principle that I heard engineers devalued as "intellectual dwarves."

But it was precisely the direction of the value hierarchy that was contested; engineers typically inverted the relative valorizations made by scientists and asserted the superiority of pragmatism over dreaming, thus evading the inferior status assigned to them by scientists. They talked about working with scientists who "get off on idiotic things, talking about philosophy," rather than attending to the important matters at hand. They viewed many of the scientists' ideas for mission scenarios as "exotic requests" that did not take into account the realities of technology or of funding. As one ESTEC engineer explained, "You can't be too intrigued in the stars all around you, or else you will never get anything done." To get things done, in short, people had to take the superior, pragmatic attitude; they had to include constraints as part of technical work. Often, the scientists concurred with this perspective; one SSD scientist, noting that most ESTEC engineers viewed the scientific community as "a big kindergarten," agreed: "Scientists behave in a very childish way because they always want to do whatever they want"—"whatever they want" was indicative of unrealistic dreaming, not of the mature compromises of pragmatism. When pragmatism was thus assessed as the superior principle, engineers emerged at the top of the social hierarchy because of their embodiment of this value, while scientists were designated as inferior. It was here that I found engineers esteemed as those who "really know what they're up to."

Reorientation

The evasive mechanism of reorientation shares features with both deflection and inversion. As in the case of deflection, in the discursive moments of reorientation people do not contest the direction of a particular value hierarchy; as in the case of inversion, people do not contest the relevance of the designated social field. Instead, people accept the terms of the debate but deny that the categories ascribed to them in fact apply, thus affirming shared values while expressing distinct interests. To take again the example of dreaming versus pragmatism: in reorientation, for instance, engineers might agree that dreaming is better, but insist that they do, in fact, have the capacity to dream. Similarly, scientists might agree that pragmatism is better, but insist that they do, in fact, often exercise pragmatic principles in their work.

To illuminate the way in which reorientation manifests itself, I explore here the reactions of scientists and engineers in ESTEC to a seminar I gave in SSD toward the end of my field stay. In this seminar, I sketched out the contours of the dreaming/pragmatism distinction, noting that it was typically scientists who were conceived as dreamers, and engineers as pragmatists. I likened their relationship to that between Don Quixote (the noble dreamer) and Sancho Panza (the worldly pragmatist):

> Dreamers are like Don Quixote, indulging in fantasy, not living up to the demands of adulthood, and the hard realities of life, tilting at windmills. Pragmatists, on the other hand, understand what they are seeing, know it for what it is, and recognize that dreams have to be given up in return for "maturity." Like Sancho Panza, the pragmatists journey at the side of the dreamers, serving them, guiding them, and making sure that the dreamers do not get involved in battles that they cannot win.

Immediately following my presentation, a scientist raised his hand and said, "I must protest your description of scientists as dreamers." Although I assured him I did not mean to suggest that scientists were sitting in their offices, asleep at their desks, he was not convinced. Later, in a private conversation, he insisted on his point: "When you define the energy range of an instrument, you don't dream." Other SSD scientists, in informal conversations with me following the presentation, also insisted that they were as pragmatic as anyone else in mission work; they repeated the claim I had heard throughout the year that they were in fact much more pragmatic than their academic counterparts, whom they always had to keep in line, not letting them dream about unfeasible scenarios and technical options (see Zabusky 1992 for details on this cultural comparison).

In this discourse, SSD scientists attempted to reorient the positions of participants in this social field and so evaded their own relegation to inferior ones. They did this by accepting that pragmatism was better than dreaming for those working on ESA space science missions, but insisting that there were different kinds of scientists. That is, they rejected the definition of scientist as dreamer (which in inversion was accepted) and argued that they were pragmatic scientists. They effected this reorientation by shifting the focus of the relevant opposition from "ESA engineer versus SSD engineer" to "SSD scientist versus those dreaming scientists in universities."

The engineers attending the seminar, for their part, heard in my description of the two positions a more positive valorization of dreaming. In private conversation following the seminar, one ESTEC engineer made sure to let me know that engineers working on scientific missions themselves "need vision and foresight, because we have to be able to imagine what we will be doing in the year 2015." Matching the reorienting efforts of the SSD scientists, this engineer evaded the suggestion that he and his colleagues might be inferior pragmatists, showing me that engineers could be as visionary as scientists.

The examples I have provided of the three mechanisms at work in the system of evasions show only one moment in what is an ongoing process of negotiation. My expository presentation here, then, itself momentarily stills the temporal process that is working together, and so may give the impression that the kinds of oppositions and resistances I have identified work in one direction, through a particular opposition in a specific context. But this is far from the truth; in practice, people keep everything moving by never settling on any one set of contradictions for long.

For instance, in the case of the administrator at Headquarters, my depiction of our conversation suggests that she was speaking in a context in which the assessments of technical people dominate, thus instigating her deflecting maneuver. But I could as easily have begun the example from the vantage point of Headquarters, where the assessments of lawyers and politicians dominate and the activities of scientists and engineers are correspondingly devalued. From such a vantage point, scientists' depictions of inferior administrators would appear to be the evasive response to an imposition of a value hierarchy generated in the social field of macropolitical cooperation. In fact, then, it is clear that the Headquarters administrator and the SSD technologists evade each other's impositions over and over again. The situation is similar for the cases I presented of inversion and reorientation. For instance, immediately after the scientist in the last example insisted to me that "we don't dream when designing instruments," he expressed a different point of view, articulating his conviction that in fact scientists are dreamers, because dreaming is better: "Without scientists, missions wouldn't be successful, because they are the architects, the designers"—in short, the dreamers.

It is thus not possible to locate a definitive starting point for this kind of ongoing negotiating; every expression of superiority is matched by an evasive maneuver, which is itself an expression of superiority, thus generating yet another evasive maneuver, and so on. Inherent in the system of evasion, then, is another source of the dynamism of working together, as people negotiate their way through the diversity that constitutes their practice. This ongoing negotiation of diversity is a critical component in the effort to articulate and defend the value of equality.

The Value of Equality

Negotiating and so undermining the assertion of the hierarchies of particular value systems is one way that participants try to maintain the value of equality in practice. But participants also talk explicitly about their resistance to the idea of social hierarchy. That is, not only do people evade conscription into others' value systems indirectly through the evasive mechanisms I have described, but they also directly deny that their own work is in any way tainted by "hierarchy." Participants name equality in their talk, expressing their perception of it as a fundamental value of their social system.

People wanted to be recognized for their commitment to equality. Participants talked about the fact that they did not like hierarchy, that they preferred not to engage in hierarchical practices, that it was only other (inferior) people who were committed to hierarchy. For instance, everyone working together on ESA science missions was fond of describing the way in which they, in Europe, worked in an egalitarian way, as opposed to the way their American counterparts in NASA worked. One SSD scientist, telling his colleagues about a meeting he had recently attended in Moscow to discuss a new science mission, noted that "there was a large delegation of scientists there from the U.S., who were well controlled by NASA; so when they were told to sit down, they did."[6] Here, hierarchy was pushed entirely outside of working together; this rendered all participants in ESA missions equally superior in their commitment to equality, a fact of which people were proud.

Some participants, however, at other times ascribed hierarchy to other people who were involved in working together, and in so doing designated these others as inferior. For instance, scientists working on

ESA science missions often suggested that this was one of the fundamental distinctions between them and engineers; whereas they themselves worked according to egalitarian principles, engineers did not. One project scientist described the way in which engineers on Project Teams accept hierarchy: "When the project manager says jump, they [the engineers] jump, whereas the project scientist can't do that with the Science Team." While engineers "just accept a decision that comes down," scientists "tend to think more about the morality of a thing and fight a decision rather than just obey the rules." In the view of SSD scientists, hierarchy emanated in particular from the ESA bureaucracy, and they talked about that bureaucracy as if it were something foreign to them. In these moments, they affiliated themselves instead with "the scientific community," which had, from their perspective, eliminated such arbitrary social hierarchies. Their academic colleagues would, for the most part, agree.[7]

ESTEC engineers, however, certainly did not locate hierarchy within their own working relations; they, like the scientists, found themselves superior in their principled egalitarianism. To assert this, though, they did not oppose themselves to scientists but to a different kind of engineer, the industry engineer. One project manager, refusing to accept the hierarchical ascription of the SSD scientists, described the Project Team not as "more hierarchical," but instead as "more organized" and "more disciplined": "Our job descriptions specify only *what* must be done, but never *how* it must be done." Other ESTEC engineers also talked to me about how "flat" they found the hierarchy at ESA when it came to really getting their work done; from their perspective, although hierarchy might appear on paper (in organizational charts, and so on), in practice, this hierarchy disappeared. In fact, ESTEC engineers did not perceive divisions based on principles of social control in their teamwork; they put the emphasis instead on solidarity, suggesting that everyone depended on everyone else in order to insure the whole. "Teamwork" was, in this way, an essential component in working together, one based on the individual initiative of equal participants and not on bureaucratic regulations. By contrast, they viewed industry engineers as working in contexts where the bureaucratic hierarchy weighed heavily, establishing and maintaining inequalities not only on paper but also in practice.

Although people are always dodging hierarchy, both by exercising the mechanisms of evasion I have described above and by explicitly denying its applicability to them, hierarchy is not universally condemned. Most people, in fact, firmly believed that some measure of hierarchical organization was necessary at certain moments—even if they also valued egalitarianism at others—if they were successfully to meet their goals of producing high-quality, well-functioning scientific missions. The constant effort to push hierarchy out of the way, to evade its orders and to denigrate its presence, is indicative, thus, not of the desire to eliminate hierarchy altogether; it reflects instead participants' efforts to control the principle of hierarchy itself, in order to prevent it from taking over their social system. In other words, participants acceded to the necessity of a practical-technical hierarchy but endeavored to keep it in its place, resisting the cultural judgments of social hierarchies that always threatened to appear in its wake.

In working together, arguments about and evasions of hierarchy are part of a struggle for social control; arguments about who can be identified with superior values, including that of egalitarianism, are arguments about who has the right to exercise legitimate authority and so control the resources and actions of others. In the next section, I explore the way in which participants endeavor to control the spread of hierarchy and the exercise of control through expressions of and insistence on "responsibility."

Responsibility and the Circulation of Authority

> I wouldn't like to be forced to do something, but to arrive at decisions this way, through consensus, is a painstaking process.
>
> *(An academic scientist, member of an instrument consortium)*

> Rule by consensus won't work because there will always be a dissenting vote. What you need to do is have open discussion followed by a decision which is taken rationally.
>
> *(An ESTEC engineer)*

Greenwood (1988: 64), writing about members of the Mondragón industrial cooperative in Spain, observed that there people "worry about hierarchy in the system, not because hierarchy itself is unacceptable, but because it is easy for hierarchy to become authoritarian rather than solidary." This is the case as well for people working together on ESA space science missions. In the view of participants, the process of working together in diversity and through egalitarian relations is one in which there must be winners and winners, not the winners and losers who appear in hierarchical systems.[8] In the realm of values, people insure this by avoiding the implications of hierarchy, as I discussed above. In the practical realm, however, people recognize that in working together, from time to time (and there are many such times), hierarchy is a necessary part of the organization and accomplishment of tasks; furthermore, this accomplishment depends on the exercising of authority if anything is ever to get done. Put another way, in order to produce the technology, to realize the scientific goals of the mission, and to organize the social groups involved in the work at hand, somebody must lead the way and, especially, make decisions.

As I argued above, decisions are dangerous moments in part because they make a social hierarchy manifest; such hierarchy is dangerous because it runs along the precipice of domination. That is, although a certain kind of hierarchy is acceptable, even necessary, in working together—not least because it insures the connection of the parts to the whole—there is always the threat that in the moment when such principles of organization are articulated (i.e., in decisions), a neutral hierarchy of function will be transformed into the judgmental hierarchy of persons. Instead of relations of equality, in other words, people worry that they will find relations of dominance and subordination. For participants, living and working in accordance with hierarchy is thus unacceptable, because the social order it potentially describes is one in which obedience to a controlling, dominating authority renders everyone the same. By giving up their ability to volunteer, they give up their individuality. The resulting homogeneity, although it offers a kind of equality, is anathema because it eradicates the diversity on which working together depends.

Debates about who has authority and how authority can be exercised in decisions are not, then, simply conversations about how to organize the social relations of technical work; they are fundamentally discussions about the social good, about how to achieve a just social order. The problem is one of governance, and people talked about this problem in explicitly political terms, as they identified egalitarianism with democracy and hierarchy with dictatorship. To participants, the hierarchy that appears in dictatorship is unjust because the political authority exercised through it is immovable, uncontestable, and exclusive: only those at the top of the political hierarchy can make decisions, the location of the top itself is not to be questioned, and the decisions that are handed down must be obeyed. By contrast, in democracy political authority is changing, negotiated, and inclusive: during a specified period of time, some people can reside at the top of a temporary hierarchy and so make decisions, but they do so always on behalf of others whose points of view are included. This means that decisions can be, and routinely are, contested. There is a fluidity built into democratic systems, in the view of participants, a fluidity that allows people to maintain their diversity.[9]

The conundrum in decisions, in sum, is this: the exercise of authority is necessary in making decisions on technical, scientific, and social matters, but the exercise of authority is dangerous because it has the potential to become domination. Thus, people struggle against hierarchy even though they need it, not because hierarchy is intrinsically bad, but because if the inequalities of a permanent social hierarchy are established, they might eradicate the diversity of working together. A hierarchy of people, as distinct from tasks, can result in this conformity because the only differences that would remain would be those between superiors and inferiors, rather than the multiple differences inherent in the equal and independent interests of working together.

People carry out their struggle against the establishment of a permanent social hierarchy in two ways. One, they insist on the value and importance of "responsibility." This is a kind of passive authority that enables participants to get things done, but through egalitarian and consensual means. Two, they insure that no one person or group acquires the exclusive right to exercise authority and so control others by resisting assertions of authority; in so doing, they keep authority circulating.

Responsibility

Early in my stay at ESTEC, an SSD scientist said to me very seriously, "That's a big word in ESA, responsibility," but he declined to elaborate. Over time, I listened to many other people also talking about responsibility in working together. Typical pronouncements included this one by a project manager, who told me that "in project work, to get work done you have to carefully delineate responsibility," and this one by another ESTEC engineer: "To make a success, you have to *share* responsibility."

Responsibility was a big word in ESA because it is what stood in for centralized authority. All participants must know what their responsibilities are, and they must also take responsibility; this is the only way that work can be accomplished since there was no embracing organizational structure in which all members were subsumed and through which people could dictate action. Working together through principles of egalitarian diversity required the active involvement of all participants, an involvement that was characterized by voluntary commitment (i.e., "responsibility" rather than "obedience"). Participants, indeed, depended on each other to take responsibility for their actions, because without this the mission could fail, and everyone lose. Thus, people or groups who did not meet their responsibilities were open to censure, since without this voluntary commitment to action, working together became impossible.

A project scientist, talking about the Science Team whose work he coordinated, told me a story about responsibility that demonstrates the significance of this principle for successful action. Early in the lifetime of the mission, some scientists were asked to leave this Science Team because they "didn't meet their responsibility." They were replaced with other scientists who were more dedicated. The project scientist explained the importance of meeting responsibilities: "This is important because on a project like this they have to get down to business, and they have to meet their obligations. People have to know that when you say a date, you stick to it. Only if you're dead will excuses be tolerated." A good deal of trust underlies this process that depends for its continued existence on the will and desire of myriad participants in relations of equality rather than the rules and regulations of a hierarchical order.

The counterpart of trust in this system is blame. An emphasis on "responsibility" meant that it could be difficult to identify the source of error, since anyone and everyone was implicated. I watched a scientist and a technician, conducting a laboratory test of an instrument, try to figure out why the test was not working properly. There was some confusion about where the problem lay; the technician suggested it was in the design of the test (i.e., the scientist's responsibility), while the scientist suggested it was in the setup of the equipment (the technician's responsibility). During an argument that was not entirely amicable, the scientist joked to me, "This is the typical battle of assigning blame." It is a battle because action originates from many points, rather than traveling downward or upward, as in a social system with a singular hierarchy of control.

The active participation by all members demanded by "responsibility" necessitates that decision making incorporate the will and desires of each member of a particular social group. This makes "consensus" an indispensable social mechanism in decision making. In chapter 4, I discussed consensus as a mechanism of harmony; here I am emphasizing not its capacity to connect people so much as its capacity to contain multiple perspectives. Consensus as a decision-making mechanism is crucial to the maintenance of egalitarian diversity, because in it all the diverse responsibilities are honored. People make every effort to match, balance, and intermesh their distinct interests and needs. Of course, this may not be an easy task; as I showed in chapter 4, consensus does not eliminate conflict but indeed generates it.

Consensus and participation, thus, require each other. Not to participate means disregarding one's responsibility, and forfeiting one's right to contribute to the consensual decision-making process. For instance, failing to attend a meeting meant that a person gave up his or her right to participate in whatever discussions took place during that integrating moment, and thus to be part of whatever consensus those who were present managed to construct. Having missed a meeting was not considered to be an adequate excuse for reopening issues, particularly contentious ones (in the words of the project scientist quoted above, only death was an adequate excuse); lack of attendance was understood only as an abdication of responsibility. This may partly explain why the debates during meetings over the scheduling of the *next* meeting were imbued with great intensity; invariably, someone or some people would be unable to attend at the agreed-upon time, thus losing the opportunity to throw their points of view into the consensual pot.

It was in meetings especially that the participation demanded by the principle of responsibility was highlighted, because it was in meetings that decisions about technical work were often made. In meetings, for instance, the results of previous actions were discussed, and new actions were disbursed. An "action" is a task, typically a technical one, such as studying the results of a test or investigating the possibility of using a certain component. New actions emerged during the course of meeting discussions, as it became clear that a question needed to be answered or a problem dealt with. Actions are equivalent to "responsibilities," because someone or some group had to agree to carry out the task in question. Meetings were the best place for assigning actions, according to one academic scientist, because "when people get actions in front of the whole group, it somehow really serves to make them do it."

The assignment or acceptance of actions is not, however, always easy. As I have said, responsibility is a passive form of authority, and people were often correspondingly passive about volunteering to act; we saw such passivity, for instance, in chapter 4, when Nigel resisted his "responsibility" to keep talking with the computer specialists. At the same time, although the chair of the meeting (at least in the case of Science Teams and other mission Working Groups) might raise actions as things that needed doing, he or she could not assign them without the consent of participants, since this would be indicative of an "aggressive" authority, of authoritarianism.[10] I even witnessed one occasion when a PI refused to accept an action that pertained to his instrument—much to everyone's chagrin, since the action in question was to conduct a particular test to make sure that his instrument would meet ESA specifications and would not interfere with other instruments' operation.

This passivity sometimes made meetings tedious occasions in which everyone avoided taking responsibility, to the extent that, in fact, nothing could get done. At a meeting of a Science Team, where the goal was to come up with a dossier of proposed astronomical observations, one discussion leader had a particularly difficult time in getting participants to talk about their interests, or to take on the responsibility to study potential research areas. After a period of halting discussion, one member of the audience asked the discussion leader, "What astronomy are you after, here?" The scientist leading the discussion hesitated, and then said:

> Well, of course that question is addressed to the group and not to me—why don't you be the first to answer it. [Laughter; silence.] I've prepared this list of possible observations to facilitate discussion and argument and not to be the arbiter of what we do. Why don't people who speak end their comments by volunteering to do something about it.

There was scattered laughter at this last remark, and the PI piped up from the back of the room, "That's going to stop discussion." However much people prefer to work in a system in which everyone is responsible, it is in fact often difficult to get people to take responsibility. "Responsibility" thus serves not only as a mechanism for insuring that things get done practically speaking; it also provides a discourse through which people can resist others' efforts to control them.

The passivity of responsibility can extend beyond meetings and the voluntary acceptance of actions. At times, people simply did not meet their responsibilities, did not carry out the work that others were expecting them to do and depending on them to do. There were few material sanctions that could be brought to bear in such a situation; only moral pressure could be applied. Such pressure was effected sometimes through covert criticism, which took the form of quiet conversation during coffee breaks and in corridors, when the lack of performance by another participant was discussed and censured. These conversations were meant to be overheard; they generated a palpable tension that quite unequivocally communicated the group's disapproval to the offending party. There was also overt criticism, which was expressed in direct confrontation with the person or people who shirked responsibility. Sometimes these confrontations took place "offline" (that is, during a break in a meeting), but just as often they exploded during the explicitly public settings of meetings. Indeed, one scientist told me that meetings were important precisely for this reason: "They can be used to embarrass people who haven't been doing their work." Indeed, criticizing people for not meeting their responsibilities was a particularly effective means of undercutting whatever authority they might try to claim, as I show in the following extended example.

In a meeting of a Science Team, the assembled scientists were discussing the composition of a report they were writing collectively. The report was to make a recommendation to ESA regarding possible scenarios for the proposed mission that the scientists were studying. One scientist, Thomas, was angry that the current, supposedly final, draft of the report did not discuss a particular area of space science, one in which he was active, and one that he claimed was a significant arena for a high-profile segment of the "European scientific community." He was persistent in his demands that the group expand the report's recommendations to include this research area. He grew angry that no one seemed to be listening to him, and his complaints began to dominate the discussion to the extent that the other scientists became annoyed, even as the chair of the meeting, Claude, a high-level administrator in the Science Directorate, tried to calm him. The meeting began to come apart.

THOMAS: [Angry.] I disagree with the goals of this report and I think that we shouldn't participate as scientists if this document is to advise the Agency on what the Agency can do. If this is to be an internal document, then they don't need advice from the community. Otherwise, I think the report should reflect the needs of the scientific community in Europe.

CLAUDE: [Calm.] I don't agree with you. The introduction of the report clearly states the goals of this group.

DANIEL: [Interrupting, coldly angry.] Did you go to the Science Team meeting in May when we discussed this report?

[The room grew quiet and tense at this question.]

THOMAS: [Terse.] No.

DANIEL: So this may explain why you don't understand the goal of this document because there the head of the Science Directorate explained carefully what the role of our group was, and the role of this document.

THOMAS: [Somewhat chastened.] But it will be a general ESA paper, which means that it will be sent to the scientific community and the member states, so it has to reflect the needs of the scientific community.

CLAUDE: [Diplomatic.] Let us look at what the foreword says, because this sets the boundaries for what we can do. [He read the foreword aloud.]

[Thomas was still not satisfied and continued to argue for his perspective, although he did so less petulantly.]

MARC: [Exasperated.] We can't have this discussion now, because it's going backwards. Here we are at the end of our assignment and suddenly again we are trying to figure out what we have been doing. I think that the terms of reference have been clear for some time now.

[Thomas was still unwilling to relinquish his point of view and continued to argue with everyone around the table. The assembled scientists grew increasingly annoyed. Suddenly, Daniel interrupted the conversation again.]

DANIEL: [Cold.] Excuse me, Thomas, but I want to remind you of one thing. Every member of the team was supposed to provide a draft of a section of the report which was relevant to his specialty. We never received any from you, so we thought you weren't interested in participating.

[During Daniel's comment, participants around the table studiously looked away from everyone else, either into the distance or at the papers in front of them. After his remark, there was quiet, and finally Claude tried to resume the discussion.]

THOMAS: [Quiet.] I'm very sorry I couldn't come to that important meeting.

DANIEL: Nonetheless, the fact remains you didn't contribute.

This interaction provides a dramatic example of the results of failing to participate. In this case, Thomas's credibility in asserting the significance of his research specialty was undermined by his inability to live up to his responsibilities, which included attending meetings and drafting sections of the report. The time for him to argue for the incorporation of his viewpoint had passed. A consensus had emerged about the meaning and scope of the collective report during previous meetings and through the assembling of drafts, when the diverse perspectives of participants were voiced and incorporated. Such a consensus could not be undone.

But this representation of the event is too final. Even decisions arrived at through consensus must be contested; any decisions momentarily express a social hierarchy, and if that hierarchy is to be prevented from becoming established, it must be undermined in practice. In the case I just discussed, for instance, the argument created winners (Daniel and Claude, on behalf of the Science Team) and a loser (Thomas). As I argued above, however, such a situation is intolerable in working together, where people value an egalitarianism that enables everybody to be a winner. This hierarchy elevating people in the right (those who participated and met their responsibilities) above those in the wrong (those who did not participate or meet responsibilities) could not be allowed to persist—and indeed it proved to be quite temporary.

During the coffee break following the debate, two scientists in separate conversations with me evaded the consequences of authority that Thomas's concession had made manifest. One told me that in mission work, everyone took a turn at being "devil's advocate" and "this morning, it was Thomas's turn." This did not mean, the scientist thereby implied, that Thomas was always in the wrong, or even that he was necessarily wrong in this case. The other scientist I spoke with made precisely this point. He explained to me that although annoying, Thomas was absolutely right about the research specialty he claimed needed to be included; the scientist thus deflected the context of our conversation to a social field in which expertise was more relevant than participation.

Through such commentary, people kept the process of working together moving and kept everyone equal. Participants even suggested that individuals had a responsibility to speak up, even if the opinion they expressed went against the consensual grain. Nonetheless, from my analytic perspective, the public moment of censure is equally necessary. In it, participants are able to reiterate the importance of "responsibility"—and, in particular, the importance of active participation to meet responsibility—as the only means available to those working in a dispersed process whose success cannot depend on the direction and control of a centralized authority.

In the view of participants, the emphasis on responsibility was commensurate with a preference for "democracy." Wherever and whenever possible, people stressed the importance of meeting responsibilities, and of including the perspectives that diverse responsibilities represented; to them, these were key components of political democracy as well. In a way, the participants in working together share a view of democracy which resonates with that of the ancient Greeks, for whom "representative democracy" was unimaginable (Dahl 1989). Instead, democracy meant that all citizens had to attend the regularly held assemblies in order to discuss and "directly decide on the laws and decisions of policy"; furthermore, "citizen participation . . . included actively participating in the administration of the city" (ibid.: 19). This is true as well of the "citizen" participants in working together; they, too, must attend the regular meetings of Teams and Working Groups, and they, too, must participate in the project's technical work by agreeing to take actions and to carry them out. (This is the normative expectation, if not always borne out in practice.)

But the Greek ideal required a homogeneity of the members of the demos (that is, the citizen body), in order to facilitate the harmonious advancement of the public good; indeed, the citizens had to be able to agree on what the public good was (ibid.: 18). Such commonality and easy agreement is, for the participants in working together, not only impossible but undesirable. Rather, they demand and depend on diversity. For the scientists and engineers working on ESA missions, the emphasis on responsibility insures individual autonomy. It serves as a necessary antidote to the dangers of control, wherein domination brings about the annihilation of difference and so jeopardizes the very existence of the social system of working together.

Circulating Authority

Recognizing that everyone should be responsible, however, does not entirely solve the conundrum of decisions, because the fact remains that decisions repeatedly have to be made through the exercise of authority by which other people are told what to do, if not how to do it. Sometimes, indeed, the democracy of consensus appears as a dream, in contrast to the reality: "Democracy doesn't work when you're doing instrument development," said one SSD scientist, "You need someone to make decisions. Someone has to decide how many wires there will be, and where they will go, although of course if you have a good reason to disagree, you should be able to—but you can't decide these things by committee." At least one consortium PI echoed this view as well: "There's a feeling that everything should be democratic and everyone should have a vote, but I don't agree. People should have a say in how things are set up but then those who are elected should be dictatorial about deciding how things should be done."

These quotations reveal that consensus works only up to a point, the point of actual decision. Then, "you need a leader to take action," and this leader exercises authority unilaterally. The fact that at some point a leader is necessitated reveals a fundamental problem with democracy, however much it may be valued: in a democratic social order "everyone is responsible and no one is responsible," as one engineer put it, which means that no one ever acts. So although democracy is preferred to dictatorship, it, too, cannot prevail, because then nothing would ever get done.

People joked about such problems all the time. For instance, during a meeting of a Science Team when the morning's discussion had run longer than expected, the project scientist asked the members of the team to return from the lunch break promptly at 1:30 P.M. so they could complete all agenda items by the day's end. At 1:40, I returned to the meeting room to find only the project scientist and two other SSD scientists working under him (in the ESA hierarchy) seated around the table. A moment later a third SSD member arrived; he said to the project scientist jokingly, "I see that the only people who are on time are people that you control." The project scientist replied forlornly, "There are some disadvantages to working democratically, by consent. It's certainly not efficient—in fact, I wonder if there are *any* advantages to doing things this way."

Accepting the necessity of control is not, however, the same as desiring it. People reviled it, as I have indicated, because of the dangers it presented: of domination, of homogeneity, of stasis. To accommodate this "necessary evil," and to insure that "dictatorship" did not become the exclusive form of social order, people worked to keep authority circulating. The principle of egalitarian diversity mandates that no one claim definitive, complete control over anyone else; there must be a corresponding ambiguity about authority, about who has it and when, for how long and in what contexts, in order for everyone to be able to keep participating by taking responsibility. In this way, participants insure that when those moments occur when decisions must be made, everyone eventually has an opportunity to exercise authority; in this way control of moments when social hierarchy is articulated is shared and equality maintained

Authority executed as part of social hierarchies of control is kept in its place discursively, as people resist its presence. Participants referred to it witheringly as "coercion" and "imposition"; they condemned the "threats" to their independence, complained about being "forced" to act in ways not of their own independent choosing. In their resistance to this kind of authority, people pulled back from the homogeneity it generated by asserting their distinctiveness. "Even if harmonizing is a good idea, people have rights," said one PI; "in principle it's nice but in practice it forces people who have been competitors all their lives to work together, and that's why the PIs resisted the notion." Another PI, a member of the same Science Team, agreed with this assessment of the idea of "harmonizing" the diverse researchers' interests: "ESA wants to rule everything; that's why they put the mission scientists on the Science Team. They are supposed to tell us what the best science is. But we [the PIs and their consortia] will not relax and let the mission scientists propose what the best science is; we have our own ideas."

One mode of resistance was to deny the validity of the terms proposed by the source of authority (such a discursive tactic resembles the mechanism of evasion discussed above). SSD scientists did this all the time when confronted with ESA rules that they felt interfered with their ability to execute their professional tasks. Take, for example, their interpretation of ESA's rules regarding hiring. ESA's personnel policy stipulated that staff members be hired according to a system of geographical distribution, in order to insure that the staff complement reflected the international membership of the Agency (see chapter 3). Although this personnel policy helped to provide an environment of national diversity within ESA, a diversity valued by staff members, people often rejected the policy as a manifestation of authoritarian hierarchy—it became a bureaucratic rule that interfered with the work of scientists.

When the need for additional scientific staff members arose, SSD division heads turned to European universities—or, as they referred to it, "the Community"—in their search for new expertise to bring into ESA. They were constrained in this search, however, not only by the technical requirements of the position (scientific discipline, level of experience and expertise, appropriate focus of research) but by the Agency's demands for geographical distribution; when a decision about hiring was at issue, the criterion of national diversity became a target for opprobrium. Their resistance to this rule was a resistance to what they perceived to be ESA's attempt to dominate their independent scientific activities.

A clear example of this occurred in the case of an Irish research fellow who was being considered for promotion into a permanent staff scientist position by his immediate supervisor and division head in SSD. The new position was one that seemed tailor-made to his experience and expertise; it focused on a mission with which he had become familiar during his time as a research fellow in SSD, and concerned an area of science and technology that had been part of the work he had done for his Ph.D. dissertation. Despite this background, however, the Personnel Department initially denied the appointment on the basis of the research fellow's nationality—he was Irish, and ESA already had its quota of Irish personnel; to hire him would put them 20 percent over the allowed number.

His supervisor and division head lobbied hard with Personnel and members of HQ, arguing for a waiver of the geographical rules because this scientist was uniquely qualified. They also pointed out that although the Irish were "overrepresented" on the ESA staff considered as a whole, there was, in fact, no Irish scientist in SSD. While the case was being negotiated, there was some discussion inside SSD about the situation, discussion that focused on the scientists' being forced to accept the dictates of a department (Personnel) that knew nothing about science and was enforcing a rule that undermined their independence. Sometimes, conversation turned to the inability of nonscientists to deal adequately with the "statistics of small numbers"; one SSD scientist involved in the negotiations joked during a meeting about this: "Personnel can't cope with small statistics; they tell you you have a factor of 20 percent too many Irish, and you point out that it's only one person . . ." Everyone assembled laughed at the unstated implications.

In the end, they won a compromise solution: the scientist was given a nonrenewable staff scientist contract for four years. Everyone accepted this solution, indicating that at the end of the four years, they would once again enter into battle with the bureaucracy, arguing against the imposition of an arbitrary rule that denied their scientific rights. By then, they assured me, the Irish scientist would have become an integral and crucial member of the mission working group, and his contract would be easily renewed. The SSD scientists felt that in this case they had successfully resisted the Agency's authority and so had safeguarded their interests. In their struggle for control, they also managed to shift the source of authority from ESA to the scientific community and so kept authority circulating.

The exercise of authority always threatens to turn into dictatorship. Not only do those who are asked to follow orders resist, but those who are in a position, however temporary, to give orders resist the implications of this status themselves. Meeting chairpersons and discussion leaders in particular found themselves always trying to downplay their superior roles.[11] Although they played a crucial role in meetings, guiding the debates and discussions that took place, helping to insure the expression of every available independent point of view, they were not, bureaucratically, socially, or practically speaking, the participants' superiors. They could not make unilateral decisions, nor could they take action independently even after having heard everyone out. Nonetheless, the position of "chairperson" contained the potential for authoritarianism, and so for the control of others, and therefore those who inhabited that position were suspect. They had to be careful to emphasize their roles as mediators rather than as dictators. Said one during a contentious debate, "There is no dictatorship here; if the group agrees that there should be something included then we will—no problem."

Chairpersons made every effort to signal the fact that they were speaking as equal participants and not as arbiters of decisions. Typically, chairpersons introduced their remarks by saying, "Personally, I think . . ." or "In my view," stressing that they were speaking with the participating voice of an independent actor, of the equal team member, rather than with the dominating voice of the Agency or other institutional hierarchy. The rejection of authority in social hierarchies made it difficult sometimes for people to accept leadership positions. One discussion leader tried, during a meeting of a consortium Science Team, not only to organize a research group to study a particular topic, but to get someone to lead the group. He had great difficulty, however, articulating his goal, as he slid from term to term in search of a way to ask someone to lead without suggesting that the person might have undue control: "It is important to have someone [pause] volunteer to—um—oversee our list of research topics, someone who would be interested in making a decision [pause] or having a subgroup work on this." The speaker attempted to encourage someone to take on authority voluntarily, but voluntary authority is an oxymoron in this social system.

Despite the threats of coercion, however, the differentiation inherent in the process of working together virtually insures that dictatorship cannot prevail. Because there is no single, centralized source of authority, it is difficult for anyone or any institution to command obedience without negotiating the terms. This is how authority keeps circulating; indeed, the very structures of cooperation almost require that authority circulate. One project scientist explained to me why this is:

> I think the ultimate authority for decision making rests with the money, which means the Project Team. But this is not entirely true, because the consortia have different funding sources. So the ESA Project can coerce the PIs into doing something—but this is a funny coercion because they have to be willing to be coerced. So when ESA says, "We will impose this interface on you," if the PIs can't live with it, they scream a lot. I guess there is no absolute coercion.

Although "screaming a lot" might not sound like credible resistance in the confrontation between PIs and those who controlled project resources, in fact this screaming, as the project scientist called it, is the essence of an effective resistance.[12] PIs, moreover, controlled their own funds, which were in themselves considerable. When the PIs screamed, then, they were supported by significant resources. Indeed, they could take their complaints to their own governments, which were the source of the instrument consortia funds; these governments, many of which also contribute to ESA funding—and thus, indirectly, to the funding of spacecraft development—could bring the complaints back to ESA. So the negotiating continues, because there is no single person, institution, or government clearly in charge. In this context of continued negotiation of multiple and independent resources, then, arguments are never "mere talk"; when people "scream," they often successfully shift the direction of decisions (as in the case of the SSD division head's hiring the Irish scientist) and so keep authority circulating.

Conclusion

A system of circulating authority coupled with a system of evasions helps to keep the social system of working together moving by giving participants the means to resist any and every attempt to secure social or cultural control. These practices further help to protect the diversity that is the fundamental value of the social system itself. By insisting on "equality" and "responsibility," participants continually undermine the potential for dictatorship, for the unreasonable (irrational) domination of diverse equals, which threatens to appear whenever and wherever hierarchy appears.

But the fact remains that decisions must be made; for brief moments, someone must be a dictator, even if his or her position is almost immediately undercut. And, in fact, people do make decisions, and participate in the making of decisions, but they do not do so lightly—they worry about this all the time. They worry, for instance, that others will resent them and reject their decisions because these decisions appear to have been "cooked up" by a few superiors, who then attempt to force others to accept the outcomes ("We resent having it crammed down our throats"). No one can bear to have decisions finalized: during one design review, the speakers kept insisting that "things are still being developed" and "lots of people are involved" and "it is not final in any sense," although one scientist joked that "in the end it will be final." If it is final, the process stops and whatever social hierarchy was temporarily expressed in a decision-making moment borders on becoming the established order.

Decisions are working together become "frozen": "There will be feelings of coercion if this is frozen too soon," worried some PIs during a Science Team meeting, as they complained that ESA was "forcing" them to come to a decision before they had been able to consult with the members of their consortia. In this way, when people worry about hierarchy and authoritarianism, they are also worrying about death, the death of the social system itself. In the following chapter, I explore these worries about death as I examine the existential dimension of working together.

Six

Existential Worries: Excitement and Boredom in the Experience of Working Together

When a satellite is launched, everyone is always afraid the damn thing is going to bust up; this makes the competition to participate more intense, and emotions run high.
(An SSD scientist discussing an observatory mission)

In the preceding chapters, I have shown how the structure of the division of labor comes alive in the practice of cooperation, as people invest such structures with meaning in the very act of dismantling them. To do this, I have explored the social, cultural, and political dimensions of working together on ESA space science missions, revealing the tensions between integration and dispersal, harmony and conflict, hierarchy and equality, evasion and responsibility. These tensions keep everything moving, and as I suggest in this chapter, such movement is the most fundamental quality of working together, one that participants will go to great lengths to protect because in it they find the source of life itself.

In order to examine the existential dilemmas that people working inside and through the division of labor experience, I turn my attention here to the affective dimension of these practices. Participants are, after all, emotionally involved in their daily activities, whether they are calculating mass budgets, negotiating technical problems, testing instruments, attending meetings, or writing reports. Throughout it all, they worry, complain, and laugh; they become angry, excited, and frustrated—these scientists and engineers are not the "robots" that one new secretary supposed she might find in an organization committed to high technology.

My intention here is to convey a sense of working together as a lived experience, as I chart the course of some of the key emotions that accompany and color people's work, and their working together. When I refer to emotions, I am not necessarily talking about the internal sensations or psychological states of individuals. Instead, I understand emotion as, in Lutz's (1988: 5) words, "a social rather than an individual achievement—an emergent product of social life." In this sense, as Myers (1986: 105–6) argues, emotions "are not simply reactions to what happens, but interpretations of an event." As interpretations, emotions are made public, and as such they are shared; they are cultural phenomena. The "mood" of a meeting, of a discussion in the laboratory, of lunch in the canteen—these are the shared emotional meanings that I explicate here.

Take the case of meetings, for example. These collective gatherings, so significant in the working-together process as I have shown, provided exemplary occasions for the observation and experience of two of the key attributes of working together: excitement and boredom. In meetings, exciting moments of intense exchange, heated argument, and focused interaction would be followed by boring moments characterized by long stretches of droning presentations, heavy lids, whispered chatter in the corner. Sometimes these would occur simultaneously: I might observe a core group of people debating heatedly around the conference table, while out of the corner of my eye, I would catch sight of someone else near me leafing through papers and making notes, ignoring the confrontation taking place in the same room. I experienced it this way: sometimes I would be writing madly in my notebook, unable to keep up with the action unfolding before me; then for long stretches, I would find myself drinking cup after cup of coffee in a desperate attempt to remain alert through tortuous discussions of mass budgets, thermal tests, gear sizes, and so on.

Both excitement and boredom accompany the technical activities in which people participate. As one SSD staff scientist said to me, "I know it sounds like I'm always complaining, but really I know nothing's perfect in life, and here the work's interesting and the pay's good." Indeed, in many contexts, people spoke to me about how "interesting," in fact, how "exciting" they found it to work on ESA space science missions. This excitement is offset, at different times and in different contexts, however, by experiences of "boredom"; boredom is itself an integral part of everyday activities.

Scientists talked about excitement when they talked about research; their "jolly," as one participating scientist put it, came from analyzing data and pondering how to answer questions about the origins of the solar system and of the universe, about the birth and death of stars, or the properties of cosmic rays. In the context of ESA missions, scientists got especially excited when dreaming up mission scenarios that had never been tried before, as they let their minds rove freely over all manner of possibilities in an effort to figure out how particular theoretical questions might be practically pursued. For this reason, they were also deeply interested in the technological aspects of mission development, as they threw themselves into the process of developing instruments and mission technology of high quality, technology that would help them to realize their goals of collecting the most accurate and comprehensive data possible. Indeed, it was in their continual attempts to safeguard scientific research interests in the building of technological artifacts that they became passionate.

Engineers talked about excitement when they were embroiled in the attempt to design artifacts that would, as an ESTEC engineer described it, "make these ideas [that scientists have] work." The intellectual challenge lay in trying to find new ways to do things, and in working on technology that was at "the cutting edge." One ESTEC engineer described the excitement of trying "to imagine what you will be doing in the year 2015." Another explained to me that his work for ESA was much more interesting than work he might do, for instance, in the automobile industry, because there "engineers try to design a new car, but here I *invent* a car"; he emphasized that this kind of work depended on "creativity and vision." These are the excitements of dreams.

When working on space science missions, however, both scientists and engineers also spent time handling the minutiae of technological detail and dealing with paperwork (as I showed in chapter 3). One project scientist complained that "there's a lot of boring stuff [to do]; the space business is awful because it requires endless reams of documentation, which is boring as hell, because of quality assurance issues. I suppose all this is important and useful, but I don't like it." Another SSD scientist exclaimed during a meeting of a Science Team, "The only way to get a physicist excited about what he is doing is to get some results; there is no satisfaction in paperwork." Engineers, too, found aspects of their daily routine boring. During one laboratory session I observed, a team of engineers was testing the integrated setup of computers and detectors that constituted a payload instrument; this entailed watching and waiting for something to happen. At one point, I caught the eye of an engineer who was in the midst of a generous yawn; he smiled and said to his colleague sitting nearby, "With all this sitting around, everybody yawns," and his colleague, stifling his own yawn, agreed. This is the boredom of pragmatic concerns.

Participants ascribed emotional interpretations not only to specific work-related tasks, but to the social circumstances in which they did this work. In particular, they viewed the characteristic diversity that motivated their practice as emotionally meaningful. Although cooperation, as they understood it, left participants cold, people were anything but neutral about diversity, that value which sprang from the structures of the division of labor. Participants were, indeed, angered by threats to diversity and grew passionate in their defense of it. Yet diversity is not without its dark side: the anomie that accompanies differentiation, differentiation in which people located their experiences of disconnection and alienation. In their daily practices, then, people responded to their encounter with difference with both pleasure and frustration.

In the first part of this chapter, I explore the excitement and boredom of participants' daily immersion in diversity, as differences (of national affiliation, occupation, discipline, institution) confront each other. Next, I connect such experiences to the dynamism of the working-together process itself, as I show how excitement is the encompassing experience in working together, because it emerges from the struggle that is working together itself. This existential excitement is manifested directly in the worries people repeatedly express about survival, worries that haunt those who are participating in working together on ESA space science missions.

Excitement and Boredom in Working Together

> People come to ESTEC because they are looking for the excitement of this kind of environment.
>
> *(An ESTEC engineer)*

> The atmosphere at ESTEC is basically good, but people here are boring; I even get boring myself working here.
>
> *(An SSD engineer)*

For participants on ESA space science missions, it is the very encounter with and immersion in difference, which the presence of diversity entails, that is exciting. The positive challenges of diversity bring excitement to participants during moments of both connection and conflict; people are enlivened both when they share their differences in a kind of togetherness that approaches solidarity, and when their differences rub against each other in argument and competition. As I suggested above, however, this same diversity has a dark side, where the encounter with difference is accompanied by feelings of boredom and anomie. These are the difficulties that come from diversity—differences separating people so that they can find no source of connection or even reason to argue, but are instead overwhelmed by indifference and apathy. As I discussed in chapter 1 and explored in chapter 3, meaningful diversity has two particularly salient forms for participants on ESA space science missions: the occupational and the national. I begin here by considering the affective associations with occupational differences in working together.

Occupational Differences

Participants celebrated occupational differences in working together because they were, as people explained it, the source of creativity. One young SSD research fellow spoke to me about these pleasures: "When doing science, it is when you are talking to someone who thinks in a different way, or who approaches problems with a different style, that you have one of these *aha!*'s, because you are working with someone who has a different way of expressing things, and who might suggest a new way of looking at things." Both scientists and engineers found pleasure in the course of mission development from their contact with the different interests and perspectives that these contrasted social actors hold. One project scientist recounted to me how much he enjoyed working with engineers on a project to design and develop a special kind of observatory telescope that would meet scientists' demands and yet be feasible to contrive; he told me of his "fascination" with this process, noting that "it's easy to say what you want to do but not so easy to make it happen." For engineers, it is exciting to work on scientific missions because they "are always weird," and engineers "love it, figuring out how to make it fly." One described this work process as involving the constant encounter with "surprises" from the "free-floating ideas" of scientists.

In the effort to work together, people were usually able to exploit these occupational and disciplinary differences, and sometimes they found excitement when sharing distinct outlooks. For instance, I observed one meeting where scientists and computer specialists were trying to craft a scientific software system. In the process of thrashing out their differences, those involved got caught up in working together through a technical problem in great detail. Although I could not understand the technical language they used, I found myself as absorbed by their interaction as they themselves were. The room was alive with energy as they talked back and forth, trying to find ways to accommodate each other's interests to solve the common problem before them, listening, debating, and ultimately connecting in a process of mutual exploration and discovery, each building on the other's ideas.

At the meeting's end, one of the scientists sat back and said, with great satisfaction, "This was a very constructive morning," giving voice to the pleasure he, and the others, had experienced during their focused interaction. In the excitement of the moment, people became absorbed in the problem at hand and *concentrated* on it; as they did so, the social and political distinctions so salient and significant so much of the time fell away. There was an epiphany of communion in this moment—people forgot what divided them as they experienced the sheer joy of being together and creating together.[1] This is excitement in the trajectory of harmony.

At other times, it is not sharing differences but arguing about them that is exciting. One project engineer, describing the Science Team meetings that he attended as an observer and occasional participant, emphasized the importance of conflict: "The Science Team is meant to be a friendly body, so most meetings get stopped before it gets nasty. In general, one behaves in a grown-up fashion, maybe too much so. It's important to let yourself go once in a while, and get things out in the open. Otherwise, it can get a bit boring; sometimes you need spice." People indeed relished the occasional "nastiness" that surfaced when participants were trying to protect their differences. After a Science Team meeting in which one of the PIs had been particularly vituperative and insistent, the project scientist commented to me that "it's not always easy to work with him, because he likes to play the bad boy, but it's fun, because he always does it so charmingly." Another PI observed to me, in another context, that "without competition, this work wouldn't be so much fun." This is excitement in the trajectory of conflict.

But the occupational and disciplinary differences among these actors also means that they do not always agree on what, exactly, is exciting about their mutual encounters. For instance, although both scientists and engineers participated in mission design reviews, the scientists tended to be bored by the technical level of the presentations at these meetings. One project scientist, informing me about one such design review, urged me to come for the beginning, when the project manager and other senior engineers would give the overview of the mission design and technological concept. The rest of it, he explained to me, would be "boring," because of the detailed technicalities presented—"The scientists sleep then," he said wryly. And when I attended the review, I did notice several of the scientists in the audience, if not asleep, at least uninterested in the presentations, since they seemed to be hard at work on private projects. When I asked some of the scientists afterward whether they had been able to follow all of the details presented by the engineers, they said they in fact had difficulty comprehending the presentations; the design review, they emphasized, was "very technical" and so it was "hard to pay attention." It was really something, they added, that interested engineers.

On the other hand, scientists got excited particularly after a satellite had been launched successfully and the data began to flow in. This, for them, was what they had been working for, and they were frustrated by what they perceived as the engineers' indifference toward these ultimate mission goals. Scientists often complained that engineers did not really care what happened to a mission after a launch; if a satellite "falls into the sea" or did not work in the manner predicted, scientists got the feeling that engineers were unconcerned, since their work was successfully completed once the satellite was launched. For their part, engineers insisted that it was not exactly that they were uninterested in the missions' scientific goals, but that they did not have the time or opportunity to learn about the science because of the way in which scientific missions were structured in ESA: Project Teams were dissolved upon launch and the engineers found new positions with other Project Teams. Still, one project manager I spoke with acknowledged that engineers' interest "drops off" at the point of launch because "once it's flying, it's proved that it works" and that was the major goal for the engineers. For this project manager, the participants' diversity in interests and enthusiasms was not necessarily a problem, even if people complained; it was an integral part of the process. He emphasized that although engineers "may not get excited about the electrons in nuclear fusion, scientists are not interested in the electrons in a circuit." Both excitements were necessary to complete a successful satellite, but they did not need to be held by the same people, or be equally relevant at every moment.

Scientists and engineers can also be indifferent to the work and interests of colleagues who share their professional orientation. For instance, inside SSD, the interests of scientists in different disciplines and working on different space science missions were so disparate that neighbors on a corridor knew very little about what activities occupied the others. Many people told me that I knew much more about what was going on throughout the Department, down the hall, and next door, than did any one of them, even though my training in science and technology was limited. Some insisted that the space science disciplines varied so widely from one another that any individual's grasp of another's scientific or technological activities could not be much deeper than my own. Indicative of this indifference to others' interests was the playful name that one astronomer told me he and his colleagues had conferred on the Cassini-Huygens mission to Saturn—they called it "bored of the rings" and professed incredulity that people could find topics in planetary and plasma science interesting.

The differences among space scientists also rendered intellectual interaction superficial and therefore boring. This was a problem that plagued the biweekly scientific seminar series in SSD. Speakers at these biweekly seminars came from a variety of disciplines; they were asked to address the entire Department, and so to speak in general terms but for a scientifically literate audience. The challenges of crafting such a talk almost always proved impossible for speakers to meet. Sometimes the seminars were deemed "superficial" and hence "uninteresting" by those who shared the same expertise, even though people who did not share that research focus enjoyed them; at other times, the seminars were dismissed as too "technical" and hence "boring" by those unfamiliar with the discipline, even though participants who were familiar with it found the same talks enjoyable.

Even work-related meetings with departmental colleagues could be boring. After an SSD Division meeting, one of the staff scientists commented to me that he had seen me "writing down who was sleeping or bored." Before I could respond, another of the scientists interjected, "Include me as one who was bored" and yet another agreed that "it's hard to pay attention." When I asked why this was, they explained that "75 percent of the time the topic is interesting only to the boss and one or two others, so most of the time you're bored." This difficulty with difference is equally relevant among engineers. One Project Team engineer found meetings of the team to be a waste of time because discussions were always "superficial," since the interests and activities of different members of the team were widely divergent.

Boredom is not the only negative consequence of diversity; so too is apathy, an apathy that is a response to the pressures of trying to find connection among differences. For instance, project scientists had to take into account the different interests of scientists and engineers, of ESA administrators and academic scientists, and of the different ESA establishments, and they often complained about feeling burdened by the stresses of this "ambassadorial" work (activity that is a critical part of the practice of cooperation in work structured by a division of labor; see Zabusky 1992). Project scientists talked often about their frustration in watching engineers "cutting scientific desire" in the course of mission development. One said to me that "the scientific ideas get so diluted through the engineering process that scientists feel in the end, 'Why bother?' since the mission will no longer do what they want it to do." In the long and intricate process of balancing the different interests of participants, "people lose interest, they get wiped out and demotivated," and it was difficult to retain even "a modicum of enthusiasm." It sometimes seemed easier to, in the scientists' words, "just stay home" and, in words I borrow from Voltaire, cultivate one's own garden.

Of course, "staying home" is also an essential part of working together. The division of labor mandates it; maintaining the value of diversity requires it. The same pragmatic project manager who recognized that the process required different interests similarly argued that participants needed to focus on what it was that they knew best and liked to do: "Everybody can only worry about his own area; this is necessary, because if you try to worry about everything you'll have a nervous breakdown." The process of protecting and fostering the value and material of diversity thus poses certain difficulties for participants—which manifest themselves in boredom, indifference, and apathy—even though diversity is also the source of that excitement which keeps people coming back for more.

National Differences

So far I have discussed the excitement and boredom that come from encounters with occupational differences; now I turn my attention to encounters with national differences. As I discussed in chapter 3, space science missions are international in their organization, and by and large people experienced this "internationality" in a most positive way. They found it "fun," "stimulating," and "exciting," although most were unable to articulate just exactly what it was that was so pleasurable. Said one Science Team member: "All these little differences are a stimulating aspect of this kind of work, differences in language and culture. We have no restrictions now [on our movements within Europe], and it is stimulating and fun to have this possibility of experiencing difference." Many participants not only took pleasure in the variety of national affiliations represented by colleagues, but viewed this as an asset. According to an ESTEC engineer, "having different mentalities means we're more flexible, more responsive to different behavior patterns," and accordingly more creative because they would be able to take advantage of whatever technical or social insights might come from having such differences converge.

It was particularly at ESTEC that I observed the daily excitement which working in an international environment brought to participants, because of the emphasis on ESTEC as a "home away from home." ESTEC was indeed sometimes described by those who worked there as "an island" or "a country"; it was a country in which having a different nationality from your neighbor was the norm (Zabusky 1993b). Even during nonwork times, many of the expatriate (i.e., non-Dutch) staff were often most comfortable staying at ESTEC and playing there, either physically or socially, because they shared the experience of national diversity with their colleagues. Such nonwork events included lunch, coffee breaks, birthday and Christmas parties, and recreation activities after work hours, including sports and dramatics.[2]

One had only to walk into the canteen to feel the excitement that was focused on national diversity. Here were hundreds of people of different nationalities assembled in one place, talking in a multitude of languages. Conversations were rich in the details of difference, and people enjoyed exploring their national differences in food, wine, and coffee as they ate their meals together. Such explorations often extended to other differences of custom, or variations in political or legal practices in different countries (Zabusky 1993a). ESTEC staff expressed the feeling of excitement they had in working in such an international environment by comparing ESTEC with national industrial firms or universities, where work was perceived to be more "boring." A departing SSD scientist worried about this: "It's somehow great to work for an institution which unites so many different countries. I don't know how it will be when I go back to [my home country]; I'm afraid it will be boring after this international environment." Part of what people would find "boring," they explained, was the way that, in national institutes, everything was "the same"—the same language, the same clothing, the same customs. It was the very presence of diversity and the opportunities it offered for immersion in difference that evoked excitement.

But this same diversity of nationality had negative consequences as well. As in the case of occupational diversity, such differences could isolate and overwhelm participants. For instance, those same exploratory conversations in the canteen that appeared to be profound moments of sharing typically masked an experience of superficiality. Despite the richness of detail, such conversations in fact frequently took the shape of a dull exercise in comparison/contrast, as people listed trivial attributes without any real give-and-take or analysis. I was involved in many of these conversations, on such topics as: different national driving laws and styles; different national rates of taxation and health insurance premiums and coverage; different attitudes toward beer, wine, and hard liquor; different political systems and modes of conducting electoral politics; and different university systems and policies for student support (in Zabusky 1993a, I offer a more extensive discussion of the contours of such conversations). These discussions provided little opportunity for a deep sharing of intimacies; participants acted instead as if they were providing official travel guides to national differences. In effect, such conversations about national differences served to keep people apart by emphasizing what divided them. Not surprisingly, participants often found such conversations to be boring.

Participants encountered the exciting and boring aspects of national differences not only in conversations about such differences, but also in the simple act of consuming food. Eating food together provided opportunities for people to affirm and remark on the different tastes associated with different nationalities. These "tastes," as Bourdieu (1984) argues, are markers of "distinction" that both identify individuals and judge them; accordingly, during meals at ESTEC, participants simultaneously shared national tastes and distinguished themselves according to their (distinct) national tastes. In both modalities they found experiences of excitement and boredom.

Nowhere was this more apparent than in the daily routine of coffee drinking. Part of the routine after lunch at ESTEC was to sit over a leisurely cup of coffee (or, in some instances, tea). In this routine, types of coffee were marked according to nationality; indeed, the coffee bar in the canteen tried to meet different national "tastes" in coffee by providing a variety of preparations, from espresso to cappuccino to regular black coffee. Moreover, cultural expectations about what tastes went with which nationality were widely shared: everyone talked about the way in which Dutch coffee was not as good as French or Italian coffee, and how it was different, too, from German coffee, although all of these coffees were better than American coffee, which was, after all, "more like water." In this way, the type of coffee someone chose announced that person's nationality, and as people sipped their coffee, individuals often remarked on these national tastes. For instance, people commented that French and Italian nationals "never" drank coffee with milk after the breakfast meal and agreed similarly that Germans and Scandinavians "never" drank cappuccino. Despite these shared, rather stereotypical assumptions about taste, however, coffee selection also provided opportunities for playfulness. People actively enjoyed violating categorical expectations in their coffee selections, and when they did, that is, when people of the "wrong" nationality did select cappuccino or white coffee at lunch, the transgression was immediately noticed and commented on, typically in joking fashion.

In this way, through the simple act of consuming coffee, the pleasures of diversity were foregrounded, as people both rehearsed and manipulated national identity in their elemental tastes. Yet boredom was not far from such pleasures; in the mixing and matching of national "tastes" in the canteen, the meaninglessness of such differences also became manifest. After all, Dutch coffee was what everyone was drinking at ESTEC, since they were in the Netherlands, even if the style of preparation mimicked other national tastes. In coffee drinking at ESTEC, participants thus also experienced the dilution of distinctive tastes, a dilution that threatened them with the boredom of "sameness" they preferred to reject.

In addition to superficial conversation and the blandness of taste, there were other, more serious, difficulties associated with maintaining diversity. In order for ESA to be able to provide "ambassadors" who can circulate among all the participants, some people must leave their homes for more than just a few days at a time—the international staff members leave their native countries to take up residence in a foreign country (in the case of ESTEC, in the Netherlands) for years on end. Working together thus requires relocation.

Despite the easy rhetoric of "free movement of persons" that is part of Europe's Single Market, it is apparent from the experience of those at ESTEC that such movement is free only in respect to legalities; it is expensive in terms of the emotional stresses and strains of relocating not only things but family and outlook.[3] Although some participants found their experiences of living in a foreign country to be exciting, rich in rewards, keeping them vital and enthusiastic in their lives as in their work, many who lived abroad experienced a deep sense of anomie. They felt disconnected from their roots, from the people and things that nurtured them and created them. They felt themselves to be outsiders, never comfortable in the society in which they made their homes.

There was indeed a great personal and social cost associated with following one's professional interests, a cost made clear in the way that people talked about the difficult transitions and decisions they and their families had faced when they came to live in the Netherlands (Zabusky 1993b). For instance, I heard many stories about failed marriages, when a spouse (usually a woman) could not accommodate to the new situation, left alone to cope every day without friends or family nearby to call on, some of them uprooted from their jobs at home but unable to work in a professional capacity in the Netherlands.[4] People talked about how difficult it was to become part of Dutch society—which was one reason they turned to ESTEC for recreation—and yet how, after years of living in the Netherlands, the thought of going back "home" was as uncomfortable as the thought of staying put. They no longer felt themselves to be "French" or "English," even if they did not feel "Dutch." Some said they were "European," but this did not offer them any help in deciding where to retire after twenty years or more away from "home."

Maintaining national diversity in working together thus exacts a price from some participants. People who must live with it all the time, as strangers in a strange land, often are overwhelmed by feelings of anomie and depression, and do not feel excited by the differences that they negotiate in their daily lives.

Worrying about Survival: Working Together between Life and Death

> He still dreams he will analyze some samples
> when he's 65 or 70; perhaps that keeps
> us alive, having such dreams.
> *(An SSD scientist, referring to a Science Team member)*

Excitement is not simply the symmetric opposite of boredom; it is the encompassing experience of working together for most participants. Boredom and indifference are minor annoyances; people complain, but, in the words of the scientist quoted above, "nothing in life is perfect." And at least in working together, things are more often exciting than not. Excitement is, in fact, the defining experience, not only because it is generated by diversity qua diversity, but because it emerges from the consequence of diversity as a social principle and a cultural value—namely, from struggle.

As I have discussed, diversity is the value that emerges from work undertaken in circumstances of the division of labor, and in the preceding chapters, I have shown how diversity generates movement in the practices of cooperation. It is diversity that propels people both together and apart in a perpetually oscillating cycle of interaction; it is diversity that urges people to seek both harmony and conflict; it is diversity that requires a commitment to a value of equality and concomitant efforts both to evade authority and to make it circulate. The tensions I have described here are often experienced by participants as part of a larger struggle to achieve connection both technically and socially. In this struggle, participants know that they are alive because neither the dissolution of anomie nor the stasis of unity has been achieved. The struggle itself is dynamic and exciting, even if it is also terrifying; this is why I argue that excitement is the encompassing emotion of working together—it is indicative of life itself.

It is not easy to work together. It entails a great deal of effort, as people embroil themselves in a variety of practices that keep them coming together, bridging differences, maintaining equality, and, most importantly, fostering diversity. People keep working together in part because they share the goal of producing an integrated mission: payload instruments must be integrated with spacecraft technology, engineering issues with scientific concerns, ESA requirements with community needs. In producing such artifacts, they become committed to keeping the process in which they are entangled moving, which is to say, alive.

But there are always dangers. The difficulties of boredom, indifference, apathy, and depression are reminders of the struggle that confronts everyone, but they are not in themselves the most worrisome dangers. The ultimate danger is death itself—death of the social system, represented by stasis, established social hierarchy, and homogeneity. It is these dangers that keep people struggling to come together through and in terms of diversity; indeed, the survival of the social system itself depends on people making these efforts to come together, even if union is not achieved. Excitement emerges from this struggle to keep the process of working together going. It is only when people stop struggling that the threat of death appears; to be working together is to be alive. The existential meaning of the experience of excitement now becomes clear: excitement is being connected, dynamic, and alive; boredom, by contrast, is being frozen, isolated—in essence, being dead.[5]

Working together is thus not only a technical process but an existential one. Even as people are engaged in its complex practices and technical details, they are always worrying about survival, because in working together, survival is a question. It is a forthright and straightforward question asked of the technology: Will it survive? How long will its lifetime be? How can we help it survive in the harsh and unrelenting environment of space? Will it survive long enough for us to learn something? Survival is also a question for participants themselves: How long will my lifetime as a scientist/engineer be? How can I help myself to survive in the harsh and unrelenting environment of conflicting interests? Will I survive long enough to produce something? Working together appears, then, as a struggle between life and death; in it, everyone is worrying about how to stay alive, and in so doing, about how to keep the process itself alive.

Worries about Time

Time was a source of worry in working together because it was a resource directly implicated in the possibility of survival. Participants thought about time constantly; more precisely, they worried about it, particularly because there never seemed to be enough of it. Sometimes their worries appeared in complaints, other times in jokes, as in the following example:

[Scene: A staff meeting of an SSD Division.]

BOSS: Personnel is sponsoring two training courses, and I would like to know if any of you would be interested in going. The first one is on "how to manage one's time."

[General laughter from the scientists, engineers, and technicians assembled.]

SCIENTIST: Is this for people who have too much time?

TECHNICIAN: I think I would like to go there sometime before I retire.

[More laughter.]

BOSS: Okay, the second course is on how to use the new computer system we are putting in the Division. Perhaps you would like to go, Pierre?

PIERRE [engineer]: [Jokingly.] I have no time.

[Laughter.]

TECHNICIAN: You should go to the first course, then, Pierre.

A preoccupation with time is not peculiar to people working on ESA space science missions; Dubinskas (1988) writes about the importance of time in different communities of "expert practitioners" in high technology organizations. He notes that "within each community, the patterning of time is a central aspect of social order and process, as well as a focal point of meaning and knowledge production" (ibid.: 4). An illustration of the way in which such "patterning" takes culturally specific form for particular social groups can be found in Traweek's (1988) study of experimental high energy physicists, where she analyzes the central role played by time in both the social relations and the cosmological conceptions of this community. Traweek shows that "in the course of a career a physicist learns the insignificance of the past, the fear of having too little time in the present, and anxiety about obsolescence in the face of a too rapidly advancing future. . . . [These are] everyday anxieties about the terrible loss of time" (ibid.: 17). She links such "everyday" anxieties, which relate particularly to the physicists' social and professional world, with particle physics' "cosmological vision [of time] that transcends change and mortality" (ibid.).

Time was also an important concern in the "cosmological visions" of the scientists who do research in the different disciplines constituting the "space sciences." Astronomers talked about their research as "time travel," because the light they observed comes from sources that are distant not only in space but in time; the study of galaxies, stars, and quasars provided information about what the universe was like when it first emerged from the Big Bang. Moreover, stars themselves were viewed as having "life cycles"; they were born, they matured, and then they died—often spectacularly, as in supernovae. Planetary scientists, too, talked about their research as a kind of "time travel," because the planetary bodies to which they sent probes "right now . . . preserve conditions which resemble those of Earth during the beginning of its lifetime," as one academic scientist (a member of a Science Team) put it. Asteroids and comets were studied not only for the information they could provide about the nature of the contemporary solar system, but also because, in the words of an SSD scientist, they were "primitive objects, frozen in time; they are considered to be remnants of this primordial accretion [that formed the solar system], and so they are capable of showing the process of the origin of the universe."

But the preoccupation with time that focused so much energy in working together cannot be explained entirely by an analysis of the questions asked in such scholarly pursuits. There was after all no single discipline analogous to "particle physics" that corresponded to "space science." This term was a bureaucratic convenience that corralled together a variety of disciplines with discrete intellectual contours. Scientists working on ESA missions had widely different research interests, foci, and methods of inquiry, as I indicated in chapter 3. Furthermore, as I have previously suggested, research itself played only a small part in mission development, which was the focus of working together. During mission development, scientists were not the only ones who worried about time—such worries extended to engineers and technicians as well, for whom the intellectual puzzles of scientific disciplines were largely irrelevant. The "everyday anxieties" about time that afflicted space scientists and engineers working on ESA missions are, then, connected less to a common "cosmological vision" than to the social system of working together in which they were all equally involved. Such worries are linked to people's daily practices and appear in the execution of the mundane activities associated with developing space science missions.

Time, as I noted, was a critical resource in working together. All the participants in working together I knew were always busy. It was often difficult for people to find time, or make time, to talk to me; I conscientiously annotated and consulted a calendar in order to keep track of complex arrangements, often made long in advance, for interview appointments and the times of meetings I wanted to attend. In fact, keeping track of time was a constant activity for everyone. Calendars were ubiquitous. People posted them on office walls; covered with pen marks, they charted dates of travel, meetings, deadlines, and holidays. In administrative offices, calendars tabulated people's comings and goings: who was in, and when, who was on leave, who "on mission" (i.e., on a business trip), who on vacation. As many of the calendars would reveal at a glance, meetings, in particular, swallowed up lots of time. And because every meeting generated more meetings, there seemed never to be enough time left for people to sit in their offices and accomplish the solitary tasks that were also part of their routine practices: "Either we sit in meetings, or we do the work," said one exasperated PI. As a result, it was not uncommon for people to work evenings and weekends in order to keep up with the proliferating paperwork.

Worries about time manifested themselves most concretely in terms of the mission schedule, which was the matrix of mission work. Schedules represent a linear kind of time, one with a definite vector of progress. The schedule itself is punctuated by deadlines and "milestones": dates for the delivery of the instrument thermal models, for the delivery of the final flight models, for completion of spacecraft systems, for integration of spacecraft and payload, for system design reviews and other evaluative procedures. Each of the deadlines or milestones serves as the foundation for the next, leading to the mission's successful completion—the launch of an operational satellite.[6] The project manager establishes such a schedule at the outset of the mission, balancing the requirements of ESA procedures with those of mission parameters. He accomplishes this, however, through a process of negotiating with the various constituencies involved. Once the schedule is determined, it serves as a contextualizing structure for development work (I return to a consideration of this transformation in chapter 7).

People worried about the schedule in part because their experience in working together did not fit with the neat stipulations of the designated structure. Practice was characterized not by orderly progression, but by a breathless pace, chaotic slips backward matched by frantic scrambling forward; there were stops, starts, missteps, and crises rather than even-keeled development punctuated by evaluative moments or the milestones of technological integration.[7] Such nonlinear dynamism was exciting in itself, but it worried people because the production of artifacts depended ultimately on the integration of technological components. That is, such worries about time are fundamentally worries about survival, both of technology (can we make it?) and of the social system (can we resist the structured time of the schedule in order to keep working together?).

Worries about Technology

As I have shown, during mission development it is the technology that is the focus of intense interest and concern, not the research questions the mission is designed to answer. This is understandable, since technology provides the vehicle (literally and figuratively) through which scientists can come to know nature. These missions had "lifetimes," in the language of participants—they were "born," they lived, and they "died"—and the longevity of missions and their technological realization was a constant source of worry.

Every ESA mission took into account the fact that the satellites would eventually "die," and the technology was designed accordingly.[8] Part of the engineers' task was to insure the survival of the various components and systems. Different missions posed different kinds of problems. For instance, the telescope on the Infrared Space Observatory (scheduled for launch in 1995) has to be kept cool so that the heat radiated by the instruments themselves does not interfere with the signals emitted by distant astronomical sources. The satellite's design calls for liquid helium to fill a reservoir surrounding the telescope, and this helium cannot be replenished once the mission is launched. This puts a definitive limit on the mission's duration. A different kind of problem is posed by planetary missions, such as Cassini-Huygens. The technological components and systems on this mission to Saturn must be designed to survive for several years, because of the time it takes for the space probe to travel the millions of miles from earth to the outer reaches of the solar system.

Worrying about the survival of technology was partly routinized in daily practice. One ESTEC engineer explained to me that in his development of an instrument design, he had to take into account the durability of each small component that went into making the larger instrument. Taking down one of these components from a cabinet in his office, he emphasized: "This tiny detector, the size of a half dollar, costs sixteen thousand dollars, and it is very delicate, easily destroyed. Because they are so delicate and so crucial, every time I order detectors, I have to order spares as well." Furthermore, he went on to explain, these parts must be assembled and stored in "clean rooms," rooms in which the air is filtered to limit the presence of dust and other debilitating effects of the terrestrial atmosphere. In ordering spares and in treating all these pieces carefully, engineers make their concern about survival part of their daily practice.

Some rockets and satellites did not even have a chance to live, as an unforeseen and unnoticed error or flaw interfered with their operation, sometimes in spectacular fashion—as when rockets exploded on launch. People were always telling stories about technological failure, from short circuits to bugs in software that prevented the artifacts from performing as they were intended. Symbolic reminders of such technological death abound. For instance, in the display cases on ESTEC's main ground floor corridor, photographs of exploding rockets and pieces of the resulting debris were exhibited alongside photos and models of successful artifacts. The possibility of the death of technology is, thus, ever present and is built into the design process. The moment of launch is a particularly dangerous and volatile one, and worries about survival become palpable. I witnessed such a moment during my stay at ESTEC; the launch was that of the first American space shuttle since the 1986 *Challenger* accident.

Although it was a workday afternoon, all ESTEC staff were invited to view the launch of the space shuttle, broadcast via satellite. This event was greeted with great anticipation by engineers and scientists participating on ESA missions, since the delay in the United States' shuttle program had seriously affected ESA's space program and the progress of their work; some ESA satellites are designed for launch by the space shuttle rather than on the European heavy launcher Ariane. About twenty minutes before the scheduled launch, people began to gather around television monitors set up in the central office building's main conference room (where I watched the launch, in the company of approximately two hundred people) and in the canteen (with capacity for five hundred). A Dutch television station provided coverage of the launch, picking up its signal from the American station CNN; because the television commentary was in Dutch, an ESTEC staff member provided an English translation over the PA system, adding observations of his own about technical aspects of the launch.

At first, the people who arrived in the conference room spoke among themselves, commenting on the press kits that NASA had provided, on the technical inadequacies of the television narration, and on the decidedly "American" identity of the astronauts; but as the countdown progressed, conversation began to subside. At twenty seconds before launch, total quiet prevailed in the conference room, which was by now packed with spectators—people stood all along the sides of the room, and everyone watched the video monitors intently. I could feel the tension in the air; people seemed to be holding their breath in anticipation and in worry. The silent fear was on everyone's mind: would the launch be successful, would the shuttle survive?

Finally, the engines fired, and the shuttle lifted off the launching pad, but there was no applause. The critical point was not yet reached, the point at which the *Challenger* had exploded two and a half years before. People were transfixed by the sight of the shuttle rising into the air, until, at sixty seconds after launch, the booster rockets separated and fell away. The silence was broken as the ESTEC commentator announced that "this is a good separation of the solid rocket boosters—that's quite a relief for everyone"; there was some scattered applause, and people began noticeably to relax. Conversation resumed. After fifteen minutes, people began to disperse, as the progress of the shuttle now seemed assured. As I left the room, an SSD scientist congratulated me as an American for this successful launch. Another senior scientist commented to me with a sigh of relief that "now things can get back to normal." On this day, the technology had survived, and people could go on with their routine work developing new technology that in its turn would have to traverse the dangerous moment of launch.

Worries about Individual Survival

It was not only technology that "died" the day the *Challenger* exploded; people died too. Survival is also a problem for human beings, whose own fate is linked closely to the technology on and with which they work. Worrying about personal survival, literally and figuratively, was a common theme in the discourse of those who work on ESA space science missions. Scientists, in particular, were prone to worrying about their survival, both intellectual and physical.

Scientists' worries about intellectual survival concerned whether and how long they would be creative and productive researchers. There was a prevailing belief among scientists that "forty is a magical number"; after that age people were no longer believed to be able to produce good quality scientific work (Traweek 1988 documents a similar belief among particle physicists). People argued, indeed, that youth was an essential component to scientific creativity: scientists were "more effective when they're younger," claimed one SSD scientist. Another scientist suggested that scientific centers needed to keep bringing in new, young staff members because "a scientific laboratory needs fresh blood or else you lose vitality"; insisted a senior SSD scientist, "You have to be able to get new ideas in and to move with the times" in order to stay flexible and to be able "to adjust to a changing environment."

An explicit personnel policy applied to SSD reflected this prevailing belief in youth as the key to scientific creativity. The policy was called "rotation," which referred to the process of moving out older scientists and bringing in younger ones to replace them. SSD accomplished this by allocating, even to the so-called permanent staff, contracts of definite length. The "rotation" was apparent only from the perspective of the Department, because it assured that a constant level of new talent and creativity was maintained in-house through the supply of "young blood." The very policy of hiring scientists in a "supernumerary" capacity or as postdoctoral fellows clearly contributed to this ongoing resupply—even though officially supernumerary staff scientists and research fellows were excluded from the "rotation" policy, given that their contracts were finite and temporary from the start—since it insured that new, and young, scientific staff would regularly pass through SSD. From the perspective of the individual scientist, however, there was no rotation, but only a linear movement from youth to age, with a brief stop within the confines of SSD.[9]

Scientists, then, have lifetimes, in the same way that stars, solar systems, and technological components do. In this case, it is an intellectual lifetime—they begin as "fresh" and "vital" and end as "less effective" and "burned out." One SSD scientist described this "life cycle" of scientists as a progression from naïveté to experience in the ways of the world; it was this experience that also made it impossible for them to stay intellectually alive. In this cycle, young scientists started out motivated by "an ideal of what a real scientist should be," and they were correspondingly committed exclusively to scientific work. This was, however, an "immature" perspective, from his point of view, because it was formed not from knowledge of how things were, but from "dreams." As scientists aged and advanced in their careers, they realized that it was impossible to stay so committed. Indeed, with age, they became more and more frustrated with their attempts to stay so focused. Many experienced "scientific failure," realizing that "they are not as great as they thought they were"; this led them to turn away from one-sided devotion to work and to concentrate on family life instead, which made them even less able to produce good science. Another significant frustration came from experience with "the institutional administration, which makes his [the scientist's] life difficult and miserable," although learning how to work within and around this administration was part of what made a scientist "mature," at the same time. Scientists turned from dreamers into pragmatists, from idealists into cynics ("I'm over forty, so I'm getting more cynical," another SSD scientist said to me). From this perspective, pragmatism represents the death of dreams that are critical to scientific research work.[10]

Scientists worried, too, about physically surviving to be able to reap the rewards of participation (i.e., data analysis) because of the length of time it could take to get a mission off the ground, and, in the case of planetary missions, because of the length of time it could take for a space probe to reach its destination. These long periods of development and flight were accompanied by the aging of participants. People (and technology) grew old, especially if delays occurred prior to launch. One SSD scientist explained to me that "a space experiment is something you do once in your life, because it takes so long: it takes you ten years in the field before you can even propose something, then another ten years before the experiment is even launched, and another ten after that before the data comes back and is analyzed, and then you're ready for retirement." This is one reason people talked about retirement often, particularly in SSD. The colleagues of the project scientist for Cassini-Huygens joked with him in 1988, upon learning of the mission's official approval, "Now you're all set till at least 2006," the date of the projected launch, and "All the meetings that will follow up the mission will take you right up to retirement"; later, in private conversation with me, this same scientist insisted, "I haven't given much thought to being old," even though he would be in his fifties by the time the probe reached Saturn, if all went according to plan.

Aging is not the greatest of their worries—there is actual physical death to worry about, too. In separate conversations with me concerning the newly approved planetary mission of Cassini, two SSD scientists commented on this challenge to survival. One said in disbelief, "It takes eight years to get there and then everybody's dead!"; the other, though more equanimous, agreed: "I am sure some will die before Cassini gets there; in fact, one member of the Science Team already died, over the summer." This death was, literally, a matter of time, the lengthy periods involved in proposing, approving, developing, and launching ESA space missions.

Some people gracefully accepted the fact that they might not survive long enough or well enough to take advantage of the missions they were developing. According to one scientist, some of his colleagues "seem to recognize that science is just a long process and you have to prepare for the next generation." Yet there were scientists for whom the notion that their ideas would benefit others and not themselves was a cause for some anxiety. During a meeting of a Science Study Team working on one of the Horizon 2000 "green dreams," the scientists were wrangling over possible mission scenarios, some wanting to design the most comprehensive and high-quality mission imaginable, while others wanted to propose something more immediately "realistic" and "feasible." One of the academic scientists finally exclaimed in exasperation, "Do we want to read about it after retirement or work on it?!" as he urged his colleagues to try to agree on the most feasible scenario, which might have a chance of being approved and developed in a short amount of time.

It is not time alone that brings personal survival into question. Survival is threatened even in the routine technical work on space science missions. Vulnerable technology—whether rockets, cryostats, or electrical wires—makes the human beings who work on it vulnerable as well, as was vividly and tragically illustrated by the *Challenger* explosion. People expressed worries about personal survival in everyday contexts of working together by trading stories of death. Scientists in particular told tales about death, tales that described people suffering in the vicinity of exploding cryostats, short-circuiting wires, and escaping radiation. These stories suggested the imminence of death, its presence near and with them as they did their work. During my fieldwork, I was present at many such "storytelling contests," as scientists vied with each other to tell the best death story they knew. On one occasion, I listened to four space scientists (three Scandinavian compatriots, one German; one SSD scientist, and three academic scientists who worked on ESA projects) as they related, over an elegant dinner in a medieval Belgian town, tale after tale about death and near-death experiences. They began by talking about their participation in sounding rocket "campaigns," which took them, and sometimes their families, to distant locations for months at a time, as the rocket was prepared and launched, and data were retrieved. One such launch campaign took place in a remote part of Sweden; one of the Scandinavian scientists told a story about a scientist, his wife, and child who made the drive out to the rocket range in treacherous weather: they became stranded in snow, and the child died. To emphasize the dangers involved in such launch campaigns, another of the scientists told a story about an engineer who "almost died" when, returning to the dormitory late at night over a dark field, he fell into a six-meter-deep well, luckily catching himself on the rim.

I remarked, half jokingly, that I had no idea that space science could be such a dangerous profession or that people were expected to give up their lives in their commitment to it. My comment was entertained half seriously and half jokingly, and other stories followed to illustrate the challenges to survival posed by working on space science missions. One scientist described an occasion on which he was working alone in the laboratory, putting together some equipment, when he was almost electrocuted; luckily the current was not strong enough to kill him, although the experience, he assured us, was unpleasant in the extreme. Two of the other scientists followed this story with their own tales of near-electrocution in laboratory settings.

Such storytelling was not always so elaborate; at times it might be limited to two or three such tales. During one lunch conversation at ESTEC, for instance, two SSD research fellows discussed the fact that astronomy, their own field of research, was a dangerous profession; in fact, one of them had recently seen a report that ranked it as the fourth most dangerous white-collar job in the world. He elaborated on this statistic with a gruesome story about an astronomer at a ground-based observatory, who, while working alone late one night, was chopped in half by the rotating observatory dome. They then traded stories of exploding cryostats, ubiquitous laboratory equipment, that demolished labs and sometimes killed people.

I, too, acquired a personal near-death tale to add to the compendium. About halfway through my field stay, I was spending the afternoon observing an astronomy research fellow in one of the SSD laboratories as he was testing a piece of equipment that was being cooled in a cryostat. At the end of the day, he and one of the technicians had to refill the cryostat with liquid helium. The research fellow wheeled the cryostat to a different laboratory on an office chair, all the while telling me about how dangerous cryostats were, and how they had to be moved with extreme caution. The two men filled the cryostat while I watched.

All of a sudden, an intense jet of what looked like steam issued forth from the mouth of the cryostat; the research fellow looked alarmed, and the technician scrambled to investigate. They exchanged ideas hurriedly, and the research fellow then turned to me and suggested that "it might be best" if I left, which I did. He later explained to me that a leak had developed in the neck of the cryostat, freezing up the O-ring that worked to seal in the liquid helium; the resulting change in pressure, as air rushed into the vacuum chamber, could have meant an explosion, but the technician, who was very experienced, thought quickly and was able to hook up a pump to equalize the pressure. The research fellow emphasized to me that this was a very dangerous situation. Later that week, we had occasion to tell this, our own near-death story, over lunch to other scientists in ESTEC.

Working on the edge of death is simultaneously exciting and terrifying. Yet although many worry about the early demise of both technology and personal careers/lives, in general people I knew were accepting of the high degree of uncertainty and risk involved in space missions, uncertainty that threatens premature death. They were accepting because they recognized that it was also possible to worry too much, in a way that could lead to cessation of activity. The European scientists and engineers saw this happen to their American colleagues in the wake of the *Challenger* accident—so overcome with depression were they that they could no longer do their work. At the same time, however, to cease to worry can also mean to cease to live, because it signals an indifference to the existential concerns that keep everyone and everything moving. In worrying about survival, people have to strike a balance that allows them to maintain the excitement of the struggle without jeopardizing the struggle itself.[11]

Conclusion

Existential worrying manifests itself in complaints about the difficulties associated with accomplishing technological and scientific tasks, about the lack of fit between the ideal (e.g., perfect technology, scientific commitment) and the real (e.g., technology breakdown, distractions and interference in scientific work). It appears in stories about time, failure, and death in which the fates of technological artifacts and human beings are inextricably linked.

Participants' worries about components and careers reflect the deep existential worry that is woven into the fabric of the social system of working together, revealing indeed that, for these people, to work is to live, and to live is to work.[12] Participants' worries about death are, in other words, worries about dying in the middle, while work remains to be done, results to be obtained. Talk about death is an expression of participants' fears of being out of control in their lives, as they see their work caught in the web of structure and controlled by others; this losing control indeed represents their alienation from the value of their own labor. At the same time, talk about death is an expression of participants' fears of losing touch with others and becoming permanently disconnected in a state of profound anomie; such anomie lurks in the structures of the division of labor where there can be no connection because there what divides people is transformed from something fluid and negotiable to something hard and immutable, keeping them forever apart.

Existential worrying, then, lies at the heart of the social system of working together, because it reveals the link between struggle and survival. To keep their work going, people must keep struggling, since the process of working together thrives only in the dynamism that such struggle entails, as people negotiate the diversity which (dis)organizes and gives meaning to their local practices. Negotiating diversity means that people must keep trying to come together and connect themselves in order to connect their technology; so it is that both expressions of solidarity and the resistance to it are life-affirming actions.

Nonetheless, the achievement of connection (integration)—however momentary—intimates the establishment of unity at the expense of difference. This makes successful production of technological artifacts and social cooperation life-threatening achievements, even though it is these very outcomes that people are striving to produce. These achievements freeze something fluid, establishing a definitive structure that threatens the unruliness of working together with a dominating unity. This is why, even though diversity poses significant difficulties for participants—boredom, indifference, apathy, depression—people defend it and foster it: because they recognize that it is the engine of the social, cultural, and political practices of working together.

Diversity, excitement, life—these are synonyms for participants in this social system, and each is jeopardized by the very success of social practices geared toward production. In the next chapter, I return to a consideration of structure, the place where this book began, as I explore how it is that people can continue to pursue the fractious and noisy practices of working together even as they produce the solid and silent artifacts of technology and cooperation. It is in this transformation of practice into structure, without the accompanying eradication of practice, that I discover the explanation for participants' denials of cooperation, those same denials which initially led me into the heart of practice.

Seven

Working Together Transformed: The Production of Technology and Cooperation

> ESA's achievements are a reminder that successful cooperation can be achieved even in an area of growing economic importance. In this sense, space is making a contribution to European unity. . . . Europe as an entity will not be achieved overnight and not all at once. It will gradually come to exist as a result of practical achievements which in turn give rise to real solidarity.
>
> *(Reimar Lüst [1987: 22], ESA director general)*

> This mission is one of the most complicated things ever put together by mankind. It has lots of bells and whistles; it is really a big technological experiment, which represents the limit of what humanity can handle. That's why no one knows everything about it, and there are always problems lying in corners waiting to be stumbled on.
>
> *(A project scientist)*

In working together, people sustain themselves through the excitement of contradiction and disagreement, circularity and negotiation, diversity and dynamism. In these practices, people come together, pull apart, argue, celebrate, resist, evade, grow enthusiastic, and fight boredom. There is no end to this process; every assertion, every action generates a response that in turn leads to another response, and so on. There seems to be no escape from these interdependent, mutually implicating practices; so it is that inside the disorderly tensions and ambiguities which constitute the social system of working together it seems impossible for anything ever to get done. Yet the fact remains that working together is a productive process and not simply a circular one. Decisions must be made, and are made; people do accomplish things. Indeed, they produce very solid artifacts: the twin structures of cooperation and technology.

In this chapter, I turn my attention to working together as a productive process and focus on the way in which, even as people continue to participate in the endless dance of negotiation and diversity, they are producing ideas and things. After all is said and done, participants can, and do, say, "We made that, and it worked." Here is the image of the Ariane rocket with its satellite payload standing on the launch pad. Such technological artifacts are not the only outcome of the working-together process. When people say, "We made that, and it worked," they are also saying, "We cooperated." Unlike technology, however, cooperation is an unintended (if, from time to time, a recognized) consequence of participants' social labors. Here are the repeated denials that cooperation has anything to do with the daily activities in which people are engaged.

The question is, how do people sustain the dynamism of practice in the face of their products' solidity? The answer lies in the participants' way of handling decisions, those quintessentially dangerous moments in working together. Decisions are, in fact, key moments in the production of technology and cooperation. As I showed in chapter 5, people resist the social implications of decisions in practice by evading the social hierarchy they manifest, but at the same time people seem to accede to the technical outcomes, which they view as necessary. Nonetheless, in order to combine production with negotiation, participants cannot allow these outcomes, of technology and cooperation, to obstruct or intrude on the fluidity of process. Instead, to sustain the dynamism of their practices, participants push the results of decisions outside of the local process entirely. Thus these outcomes are displaced from the social practices that created them, and are thereby transformed into structure.

Because the outcomes of decisions, in a sense the decisions themselves, are in this way decoupled from local practices, and specifically from those individuals (or groups) who are responsible for the making of particular decisions, they are correspondingly able to remain intact. Once transformed into structure, technology and cooperation appear to exist independently of those who produced them, and so are not subject to further negotiation. They appear instead to be part of the a priori world according to which people carry out their work. This displacement, or externalization, of technology and cooperation is an essential part of practice. It must happen if people are to survive in the noisy, contentious, and stimulating process that is working together. It must happen if the social system itself is to survive.

In what follows, I consider in turn these dual outcomes of technology and cooperation. I explore the relation each has to the practices that produced it, and I show how each serves as the medium of working together, even while each is being produced by the very practices it helps to generate.

Technology

> Of necessity you interact, because what you produce interacts.
> *(An SSD scientist)*

> Even if there are disagreements between people, they have to get along because they have to work together to make the satellite work.
> *(An ESTEC engineer)*

That technology is both the material outcome and the material medium of the working-together process seems self-evident, both to the observer and to the participants. As I have shown in the preceding chapters, the day-to-day routines of working together on ESA space science missions were focused on and practiced through the design and construction of technology. In workplaces and during work moments, it was the technology that people talked about, wrote about, argued about, and ultimately worried about. When I refer to "technology" here, I include physical equipment (such as instruments, mirrors, motors, sensors, solar arrays, and the spacecraft bus) and software systems (such as those for spacecraft operations, data handling, and data processing), as well as the technical minutiae of specifications and requirements relating to both equipment and software (such as the temperature range for optimal instrument operation, or electrical budgets—how many watts a component can consume—or even what kind of equipment a computer system must be designed to run on; see Pinch and Bijker 1987 for a theoretical consideration of the definition of "technology").

The overt importance of technology was a basic fact of life for participants. Even when pieces under development were not present physically in local environments, people kept in contact with technology by surrounding themselves with artifacts, or representations of artifacts, from previous missions. Individuals decorated their personal work spaces with emblems of technology; these served both as mementos of their involvement in previous space science missions and as forms of reflection on the motivation and intention of all working together. Many people covered their walls with colorful posters celebrating particular missions, and some put stickers representing ESA missions on their doors or on their briefcases. In SSD, one staff member had a glass cabinet in his office in which he kept samples of the detectors on which he had previously worked. Another scientist displayed the engineering diagrams of mission payload instruments with which he was actively involved though they were located in dispersed external de partments. The public spaces of the institutional environments also testified to the significance of technology. At ESTEC and other similar establishments (such as the Max Planck Institute in Germany and Saclay in France), technology was a major theme of decoration. At ESTEC, for instance, the ground floor corridor was adorned with several display cases in which were exhibited models of ESA rockets and satellites, diagrams with information about specific mission hardware, and photographs of rocket launches. Some of the conference rooms also exhibited photographs and/or models of ESA missions (rockets or satellites).

For those who are working together on space science missions, the explicit goal of their collective activities is to produce technological artifacts that function successfully—that is, the satellites must fly and send back data. Such a functioning satellite is one in which different technological components and systems work together smoothly. Although people want to insure a first-rate collaborative performance among these artifacts, they also attempt to safeguard the needs of individual pieces. In designing, developing, and manufacturing technology, participants accordingly balance the collective with the individual demands of the technology. They do this by negotiating about the specifications, requirements, and technical parameters that circumscribe the production of artifacts. This negotiation over technology forms the substance of the dynamic practices I have described in the preceding chapters; in this sense, technology is the medium by means of which people experience the social, cultural, political, and existential dimensions of working together. It is while people are arguing about wires, gaskets, detectors, temperatures, vibration tests, and interfaces that they are taking responsibility, evading domination, invoking harmony, provoking conflict, sparking excitement, and worrying about survival.

Although the practice of working together means negotiating technological parameters, once decisions have been made—whether these concern the number of wires to be included in a detector or the amount of electricity to be made available to an instrument—the "negotiated" quality of technology disappears. The outcomes of decisions become instead part of structure and so serve as external resources that aid participants in their subsequent (and ongoing) negotiations. These artifacts appear to be unmodifiable. They are recognized by participants as historic achievements; such "history" becomes the context that shapes the work in which participants are currently engaged (I discuss the meaning of "history" in more detail in the following section). For this reason, once certain properties and parameters are decided on, they can be altered only with great difficulty. To call attention to the negotiatedness of already agreed-upon decisions is to imperil the life of the artifact and hence the life of the social process itself. This is why any such attempts will be met with hostility and heightened emotion, as well as corresponding efforts to soothe through the invocations of "the same boat." Of course, such attempts are routinely made, since the determination of when a decision is actually final is itself open to negotiation. Nonetheless, inevitably such a point is reached; there comes a time in the development of technology when decisions must be accepted once and for all in order for people to take appropriate action. In the moment of taking such action, all the negotiations that went before disappear into structure.

The struggle over closure was illustrated dramatically during one Science Team meeting when a PI began to argue with one of the project engineers, questioning certain technical specifications that mandated the temperature which payload components were required to withstand in testing (the "bake-out" requirement). He insisted that the temperature in the specs was unnecessarily high, and that the reasoning behind the establishment of this limit was ill-founded. Furthermore, he claimed that this specification would adversely affect the performance of the instrument for which his consortium was responsible, and he refused to participate—that is, he refused to follow this spec.

The other PIs present during this diatribe were understandably annoyed, if also concerned. Members of their instrument consortia were already well under way testing their hardware in accordance with this specification, and making whatever modifications proved necessary. If the spec was ill-founded, every instrument was in jeopardy; on the other hand, the recalcitrant PI's refusal to participate in the joint effort, which was proceeding in terms of this spec and others, also jeopardized the health of every instrument. If even one instrument failed in this critical area—if it could not survive the maximum temperature that occurred during flight—it could interfere with the normal operation of all the others. For his part, the project engineer was furious at the suggestion that the spec had been arrived at capriciously. He assured all the PIs that a great deal of deliberation, including input from instrument consortia, underlay the establishment of this specification. The recalcitrant PI, unconvinced, continued to argue; finally the engineer exploded in anger: "I'm not reopening the specs, no way!"

In this instance, although the standard had been set, one participant made an effort to call into question the validity of a technology-outcome; this questioning resulted in the (temporary) recognition that the constraints which inhibited action today were the result of decisions made in the past in a context of negotiation. Such a recognition was intolerable for most participants—anger and hostility could be the only response to such an audacious attempt to remake history and to call structure into question. In this case, in fact, the suspicious PI was finally silenced, and members of his consortium continued to work on their instrument to make it meet the bake-out requirement. The outcome of the decision regarding the establishment of bake-out temperatures was effectively "closed"; the specification was not negotiable. Indeed, it was important that it not be recognized as having ever been negotiated. Like other outcomes, it must be allowed to maintain the (apparent) inviolability of structure or it could not provide people with the means of and context for their ongoing activities. So it is that the negotiations of practice never cease, yet artifacts are produced.[1]

One effect of this externalization and concurrent objectification of a technology that in practice is immanent and always changing is the inversion of the productive direction. Even though, on the one hand, people recognize and point to technology as the intentional outcome of their collective actions, on the other hand, many participants put the technology first, noting that it is the real "driver" of working together. In so doing, participants displace their social and subjective needs and desires onto the technology, and then turn to the technology to validate their claims that a certain action is warranted. In the language of working together, it was the technology that had "needs" and "requirements," and these were distinct from and independent of those people who negotiated the needs in the first place.[2]

People were always talking about technology in ways that demonstrated its controlling influence over their own actions. "Many institutes have to be involved in scientific experiments because nowadays the experiments are much more complex," explained one engineer. In other words, people keep coming together in the production of complex and expensive machines because the machines have many distinct pieces whose realization requires the diverse expertise of many participants, as well as the financial support they bring. This is the practical-technical explanation for the division of labor, an explanation that resonates down through and to the level of practice. In this view, the social system of working together itself depends on machines, not people.

Technology leads people in their collective practices not only by structuring them, but in their very execution. "There are instruments which need each other," explained one ESTEC engineer, adding that "therefore, people are not independent and they have to rely on one another." A project scientist described the "pressure of the instrument" that brings about social conflict because those people "in the swamp, dealing with the nitty-gritty, cannot afford to be academic-minded." That is, people (such as the recalcitrant PI in the example above) cannot elude the grip of the technological artifacts for which they are responsible. They can be trapped by an artifact that demands protection even at the expense of some "bigger picture" in which their actions may also be implicated.

In particular, in the development of scientific satellites, people accede to the technology's need to become "integrated." As I discussed in chapter 3, much of the process of technological development is oriented toward the further "integration" of more and more integrated components. In the case of scientific missions, the critical juncture is the moment at which mission payload and spacecraft are integrated; it is the uniting of these two distinct subsections that makes a given satellite scientific, because it is the payload which insures that the satellite fulfills the mission's scientific goals. Although I say "moment" here, suggesting the occasion when instruments and spacecraft are literally screwed and otherwise fused together, this integration happens over time. The parameters for integration are negotiated and tested over the years during which instruments and spacecraft are being designed and developed in order to make this specific moment possible; in short, everything is built with this in mind.

Working together thus moves steadily toward this final integration,[3] and people talked as if it were this technological necessity that dictated their social practices. It was "technological integration" that "forced" people to compromise, to find their way to agreements, to overcome or at least organize the diversity of interests that might divide them along the way. Indeed, to participants it often appeared as if without the technology, they would not talk to each other at all, would have no "need" to talk to each other. It was only because of the technology, and their commitment to it, that people worked together; there would be no social integration but for the existence of components that needed to work together.

In practice, the tendency toward social integration is actualized in response to dispersal, just as dispersal occurs in response to social integration. It is thus distinct from, although connected to, technological integration, which is the climax of mission development. The social process is commensurate with the technical one up until the moment of component-integration, when, in virtually the same moment, dispersal appears in practice—people go off again to pursue the routines of their "ordinary lives"—alongside the concrete outcome of the completed, integrated artifact. This artifact, although distinct from the practice in which it is nurtured, reflects the values and principles of the practices that produced it.

The structure of technology provides an organized and schematized version of diversity. It articulates, in essence, a silent and harmonious version of the division of labor that the practice of cooperation noisily negotiates and contests. The technology appears as a unified entity, in which the needs of all the distinct parts have been taken into account and facilitated in the composition of the whole. Unity is achieved through the technical specifications and mission goals that together have constrained the development of the payload instruments and spacecraft, and enabled their very integration to take place. Inside the unified environment of the integrated artifact, instruments and spacecraft work together in a "spirit" of commonality, a commonality located particularly at the boundaries where each connects to the other. Unity, however, does not correspond to uniformity. The commonality of the operating environment that encloses and connects the pieces incorporates but does not eradicate the distinctions among the components. Each instrument retains its idiosyncrasies and continues to work in its independent way. The pieces are, in fact, knowable only at their edges where they "interface" with all the other pieces; internally they are known only by their makers.

The integrated artifact of the scientific satellite expresses, thus, the structural principle of unity in diversity. This principle corresponds to the dialectical interplay of the various opposing forces that animate the working-together process, but in detemporalized form. Such is the nature of structure, which is, as I indicated in chapter 1, atemporal, existing outside of time and space, unlike the uneasiness of practical consciousness, which is grounded both in time and in space.

In the case at hand, the structure of technology that articulates the principle of unity in diversity also identifies unity as the encompassing value (Dumont 1986). It is unity that hierarchically organizes the diversity of the division of labor; in so doing, unity both gives diversity room and scope for action and prevents it from dissipating and fragmenting into anarchy. The power of integrated technology for participants in working together, then, comes from its ability to contain all the differences that in practice seem continually to threaten the very production of technological outcomes. This power is itself rooted in nature, in the sense that this structure of technology seems self-evident, requiring no explanation; thus, people are free to construct their social systems in technology's image, which is to say, in nature's image, without feeling threatened by the domination of human and social institutions, such as nation-states and bureaucratic organizations.

The artifacts of integrated technology are, like the totems discussed by Lévi-Strauss (1963b: 89) "good to think"; that is, they have not only practical uses ("good to eat"), but intellectual and even moral ones as well. It is through and in terms of technology that people come to understand their social process and the cultural values that they tacitly espouse. So it is that in practice, participants often look to the technology to tell them about how their social system might best be organized: in the words of a project engineer, "We have to work together, although there are real differences dividing us. But we have to make it all work together as a system." Such "rules" about practice impart value as well. Technology reveals, in essence, the natural necessity of a social order predicated on the principle of unity in diversity. The completed artifact expresses the achievement of harmony and integration that in practice eludes people. Its composition reflects this principle and value, which itself is generated by and simultaneously motivates the process of working together.

At the same time, this power of the technology-totem is also a dangerous power. The achievement of harmony is also the extermination of struggle—the containing of differences can too easily become the subordinating and even domination of differences. And it is when diversity is eradicated that survival is threatened. This threat is indeed part of what motivated the recalcitrant PI in the example above; he was led to the otherwise unthinkable strategy of calling a technology-outcome into question because he (on behalf of his instrument) felt oppressed by that outcome, even though it represented the collation of the group's diverse interests.

Working together depends for its survival on the diversity that excites the participants as they struggle to find connection in the practices that constitute their social system. Such excitement in turn can be maintained in the face of final decisions and definitive outcomes only by the externalization of the resulting technology. The secretion of technology from the fluid of working together is, in fact, a necessary and inevitable part of the struggle to stay alive. It is by means of this mechanism that people are able to produce technology, their explicit goal in working together, without bringing their social system to a halt. These technology-outcomes are in turn able to retain their structured coherence and solidity by organizing themselves according to the same principles that in practice people use continually to disorganize their social system.

Cooperation

> The conflict within the team is potentially very dangerous. I think ESA should come up with some specifications for the whole group to follow.
>
> *(A university scientist, member of an instrument consortium)*

> I'm trying to find the common ground, no matter how shaky, that we had last time.
>
> *(An SSD scientist speaking at a Science Team meeting)*

Cooperation, like technology, is a basic fact of life for participants. All around them, they see and experience evidence of its existence. In chapter 3, I discussed the concept of "cooperation" that most participants hold; here, by way of review, I offer one more comment about "cooperation" from a participant. When I asked an SSD scientist about how cooperation worked, he waved his hand—taking in the papers on his desk, the equipment in his office, indeed, the concrete and glass of the building itself—and said, "The result of trying to bring countries and people together is all this, the whole Agency interface." ESA, that reification of cooperation, makes it possible for scientists, engineers, and others to carry out their work. It gives them a place to sit, equipment to use, and problems to solve. Without cooperation, it would seem, there would be no working together at all. The parallel with the controlling force of technology is evident here.

This cooperation is an outcome of working together, as is its counterpart, technology. Sometimes, the outcome of cooperation was recognized by participants as emerging from their practices. As one project engineer put it, referring to the work of the Project Team: "Our job is not to cooperate but to drive the issue and in the process to drive it not just by bureaucratic means. . . . A natural outcome is cooperation, because it is the easiest means of getting it done." This formulation of the relationship between working together and cooperation demonstrates that although people might recognize cooperation as an effect of working together, it is not the intentional goal of those practices. It is, instead, produced as a by-product of technical effort. What is *explicitly* important in working together, as I showed in the previous section, is to produce a technological artifact for scientific use. As it happens, in realizing this goal, the diverse participants who confront each other in practice, now coalescing and now repelling each other, transform their contrary practices into unified structure. Working together is transcended in the form of cooperation.[4] This helps to explain why participants generally deny that "cooperation" is connected to their activities.

Cooperation exists in the (reified) form of agencies, nation-states, and a variety of other such social communities. In practice, people are always constructing local social groups and then undermining them because they must keep such groupings provisional to avoid stasis, evade domination, and maintain excitement. Yet on the level of structure, "communities" appear to be firmly bounded and secure loci of unity (which is not to say uniformity, as I argued in chapter 4—"community" is an organization of diversity). People work together in a social space that is in fact surrounded by and constituted in terms of such objectified social groups, but they experience and refer to them as external to their local practices. Cooperation as it is instantiated in these communities therefore becomes a resource to participants in their daily activities. Such cooperation indeed comprises a range of resources, including concrete things like money, equipment, and organizations, as well as such abstract things as values and social connection. Furthermore, these communities represent the achievement of unity, a unity that, as is the case with technology, gives diversity room to breathe in practice while putting limits on its centrifugal tendencies. So participants use them in practice also as reference points for invocations of stability and as resources that can be negotiated.

Because of their usefulness in practice, participants were certain that these "communities" existed, but they were also certain that these communities always existed somewhere else. For instance, SSD scientists talked about their relationship to one such community—the scientific community—on a regular basis. Whenever they spoke of or referred to it, it appeared as a thing "out there," as an entity that was distinctly separate from those who were inside SSD. Whenever staff scientists spoke of this community, it always sounded as if it began with a capital C. They said things like "We have to make the *Community* aware of it," or "Someone from the *Community* asked me about this," or "The *Community* is sensitive on this topic"—referring to it almost as if it were an independent state, equal in status to France or

Britain. As something "out there," it could not be simultaneously "in here," where working together took place. It was structure, not practice, and so independent of their own actions.

Cooperation, however, did not appear only on such a remote level. It also existed as the more local instantiations of agreement, group cohesion, and consensus that are directly produced in working together. That is, when a decision is made on a technical issue, the physical artifact is not the only outcome; another outcome is social integration. The artifacts themselves are the tangible representation of social groups that in practice are always coming-into-being, but never established. So it is that an integrated satellite serves also as a symbol of "cooperation"—it is "good to think" because it tells participants about the social outcome of their practices.

Meetings are especially efficacious moments for the transformation of working together into cooperation, in part because, as I discussed in chapter 4, the rhetoric and structures of cooperation contextualize these potentially integrating moments. Such collective gatherings also provide key opportunities for the expression of the value of harmony. It is this trajectory, fervently enunciated during meetings, that leads out of practice into structure. It accomplishes this by transforming *moments* of connection into *structures* of connection. This is why the structure of cooperation resonates so strongly with the idea and value of harmony, but the two—structure and trajectory—are not and cannot be the same. In working together, harmony is the *invocation* of "the same boat"; in cooperation, and its mirror technology, harmony *is* that "same boat." This is one reason that efforts to individuate and to foster conflict emerge so strongly during meetings. Assertions of conflict, the continued murmurings of diversity and independence that color presentations and interactions, are a necessary antidote to the ominous presence of cooperation that hangs over and emerges from meetings. People must struggle to undermine the very object they are unwittingly creating, in the very moment of its creation, in order to stay alive.

The transformation from practice into structure is effected in particular through the cultural mechanism of consensus, which is both an explicit goal of decision making in meetings and simultaneously part of the trajectory of harmony. Reaching a consensus usually involves protracted argument among participants; consensus is negotiated in conflict as is technology. Indeed, the *achievement* of consensus is equivalent to the technology-outcome of a decision, and it, too, is externalized.

In the case of consensus, externalization means not only that particular technical details are turned into structure, but that the social process by which the decision was made becomes enshrined as "cooperation"—decisions arrived at reveal "group cohesion." This explains the experience of people in meetings who felt that they could really "see" the groups of which they were a part. The transformation of the mechanism of consensus into the outcome of consensus means that no matter how difficult, "painstaking," and conflict-filled the process might have been, people allocate the differences that helped forge the consensus to "history."

I have put "history" in quotation marks here to signal that this is a local term, one that emerges in particular inside the trajectory of harmony, alongside consensus. "History" was invoked especially during meetings, when discussions threatened to disintegrate into a mere recital of differences and individual interests, the effort to find connection apparently forgotten, or at least ignored. It was at these points, when conflict was on the verge of sundering connection, that someone invariably announced, "Well, that's all history," or "That's old business, so let's not talk about it, let's move on." By asking people to treat something as "history" or as "old business," participants were entreating others to stop dwelling on what had happened in the past. They instead tried to lessen the tension that had resulted from past decisions, whether administrative, technical, or scientific. "History" meant: yes, we know there were disagreements, but that's in the background, that's over and done with; let's forget it and deal with what is on the table in front of us right now.

Even though history was, thus, in some ways, ever present ("It happened like this"; "No, it happened this way"), and people were conscious of how things came to be the way they were, during the working-together process history was denied. By saying "That's history," participants emphasized that it was not useful or even honorable to know what had happened before, because it meant reminding people of past conflicts that had been overcome in particular consensual instances. Reminding people of their individual interests, specifically those that had been subordinated to the collective effort, was a sure way of unraveling the structure of cooperation on which working together itself depended. So it is that, once achieved, consensus becomes part of history; more precisely, it becomes part of the structures of cooperation that are retrospectively perceived. From fractious diversity emerges the common good, and this good becomes the new, self-evident point from which to argue and around which to work.

The "common ground" is a significant local instantiation of cooperation. It represents, to paraphrase Marx ([1859] 1977), congealed working together, and as in the case of the commodities that Marx describes, it takes on a life of its own and appears to be independent of those who produced it. In particular, the contentious negotiation of diversity disappears in the crystallized form of static harmony. Participants have difficulty recognizing their connection to such orderly outcomes, as I demonstrate in the following extended example.

As I discussed in chapter 2, scientists from all over Europe compete with each other for the right to determine what scientific missions will be undertaken by ESA. Mission selection is one of the most intense moments of working together because it makes explicit the way in which competitive negotiations must be transformed into cooperative achievements. From conflict and disagreement, people are expected to commit themselves to unity and agreement. It is not only the fierceness of competition that imparts emotional intensity to mission selection, however. Selection is also a key integrative moment in working together, because it explicitly converges the work of ESA scientists and engineers with the work of scientists dispersed in (primarily European) universities and institutes, and of engineers dispersed in European industrial firms, as well as with the political bodies that govern ESA and through which decisions pertaining to policy are made. As such, mission selection (and I am focusing here on the two- or three-day sessions at which work is presented and recommendations made) highlights and reveals the diversity—and the contradictions that such diversity entails—which keeps the working-together process alive.

In October 1988, during my fieldwork, a medium-sized ("blue") mission was to be selected for formal incorporation into the Horizon 2000 mission plan from a field of five competitors. As prelude to the intensive negotiations over selection came the Phase A presentations—or the "jamboree," as the scientists in SSD liked to call it—at which the results of the mission studies were presented to an audience of inquisitorial academic scientists (approximately 250 people attended this session). It was a "jamboree" because it was an opportunity for the celebration of community (that is, of scientific community) in a context of conflict. Its goals were social and cultural rather than technical or scientific, as SSD participants explained to me in no uncertain terms, even if the official pronouncements of ESA administrators emphasized the latter.[5]

The director of the Science Programme opened the public proceedings with remarks that emphasized the equality of the diverse participants but also the hierarchical direction of selection, in that one competitor must emerge as dominant, superior to the others:

> I am pleased to see so many friendly faces . . . and I hope to see them again afterwards. [Laughter.] . . . We have to choose one from these five missions, each of which has superlative qualifications; but making choices is what sci-

> ence is all about in the end. . . . The rules of this game call for you, the scientific community, to express your opinion; your role is not small. It will be a difficult decision, but whatever choice we make, this will be an excellent choice.

After two days of deliberations following the two days of presentations, the advisory committees recommended approval of Cassini, a joint ESA-NASA mission to Saturn. At SSD, where I heard the news, there was both rejoicing and angry recrimination. I spoke with one senior scientist who had been involved with other selections, but as an external scientist, not a member of SSD. A nonpartisan (i.e., noninvolved) observer of the selection procedures, he commented to me that he was always amazed by the process because it was always so successful. By successful, he meant that the missions selected always functioned well and returned excellent results. He called mission selection "a miraculous process" and explained that "it is not at all rational, but mystical," because "a project rises from the conflicting issues and confusions like a phoenix from the ashes." Some scientists who had participated in the competition were less sanguine but also acknowledged the validity of the outcome: "In the end of the day, no one begrudges which mission came out on top; it's just the way that it's done that rankles. But there's no animosity now."

In this example, people recognized the outcome—"the phoenix," consisting of both technology and cooperation, as a mission represents both artifacts and social groups—produced by the ongoing process, but denied the conflict that was part of its production: "There's no animosity now." Significantly, participants denied such conflict not in the continuing practices (there, things continue to "rankle"), but in relation to the particular achievement. The decision to select Cassini immediately resulted in Cassini's transformation from negotiated impermanence to cooperative structure. Its selection now gave it the power to draw people together to work on it; as structure, it displaced the arguments that had made it tenuous in the past. It was now the common ground and so could provide resources for action. The scientist in this section's first epigraph, appealing to ESA for help in establishing connections among endlessly conflicting disparate groups, treats the Agency itself as just such a common ground. Yet locally, particularly in the immediate aftermath of selection, there seemed to be no such acceptance, no firm structure. People continued to argue and disagree, deconstructing the object discursively in the here and now by calling into question the process that had produced it.

This discursive undoing of the decision does *not*, however, affect the outcome of the decision, which remains intact, because it has been pushed outside of practice. Inside practice, negotiation continues, and this is, indeed, essential to the survival of working together. Participants must continue to struggle against even those structures they have helped to produce; were they to stop, to give in to the static connections of such artifacts, they would be acceding to social inequality. After all, in this case choice means "this mission is better than that one." So it is that working together can produce cooperation without the participants' recognizing their role in its creation, as they continue to activate the dynamic practices that keep them alive.

Conclusion

Cassini, like other outcomes, became part of the structures that constrain practice, structures that express the organizing principle of unity in diversity. As such, they demonstrate the achievement of harmony; this is a unity that fosters commonality while maintaining distinctions (as in "harmonization"; see chapter 2). In essence, both technology and cooperation represent the integration of disparate pieces at a higher level. Only in these structures can the user (in the case of technology) or the participant (in the case of cooperation) locate the static unity that seems to be the opposite of local struggle.

Both technology and cooperation, albeit in different ways, thus meet the challenges posed in working together by the diversity resulting from the division of labor (e.g., the challenges of dispersal, conflict, domination, and boredom). Each, as structure, provides a means of organizing contradictions, connecting diverse pieces, and so crystallizing harmony out of the disunity and ambiguities of conflict. They achieve this without entirely eradicating the sources of conflict, namely, the distinctions among people, things, institutions, and values. But because technology and cooperation are both structures, they do not promote excitement or convey movement. This makes them inherently dangerous as well.[6]

In the case of cooperation, the primary danger comes from the association of the principle of hierarchy with human institutions (including nation-states and bureaucratic agencies, as well as smaller groups such as teams and departments), and so with the people who inhabit the social hierarchies of such institutions. In such cooperative structures, differences among participants are organized and coordinated hierarchically, and with hierarchical superiority comes control. It is the ability to control others, which the definitive establishment of a particular social hierarchy brings with it, that threatens working together because it threatens people with the homogenizing effects of domination.

In these practices, as I showed in chapter 5, although people acknowledged that the practical-technical hierarchy of the division of labor, itself mandated by the technology, might be necessary for decision making, they simultaneously resisted participation in those social hierarchies which decisions made manifest, because they were potential sources of oppression. That is, articulated social hierarchies remained viable only as *temporary* solutions to the need to produce outcomes through working together. In their institutionalized forms (e.g., as cooperation in political or bureaucratic guises), on the other hand, such hierarchies persisted inappropriately, and therefore dangerously. Established social hierarchy is, in essence, a source of domination—and domination's consequence, conformity; hence, it is potentially lethal to the social system of working together itself. It carries with it the smell of death, for in the silencing of diversity it threatens the struggle that keeps working together, and its participants, alive.

The only way that this social system can survive is for people to push the unity of enduring hierarchies out of it altogether, and to deny that they have any relationship to their practices. This denial is made explicitly, but it is also made in practice as people emphasize struggle, immersing themselves in conflict apparently at every moment. On the ground, conflict seems to be everywhere and cooperation nowhere, as expressed by participants' anxious or jocular "There's no cooperation here." In fact, cooperation cannot be where people are, because it is forced, by definition, to exist above and beyond the local level. The continual emergence and experience of conflict is not, then, as participants worry, the result of a kind of moral failure to find and commit to harmony. Participants actively embrace conflict, and they are excited by it, precisely because it is the living, processual solution to domination. People choose conflict, and in so doing, they release themselves from oppression and escape from boredom into excitement. Conflict is, indeed, the fundamentally moral choice in a life committed to diversity.

Epilogue

Sacred Cooperation and the Dreams of Modernity

> The peaceful exploration of space began, for Europe, as a dream for some farsighted scientists. Today it has far outstripped their intentions, but it is our duty, and our privilege, to keep the spirit of their ambitions alive.
>
> *(Reimar Lüst [1987: 23], ESA director general)*

Cooperation Revisited

The story I have told in the preceding pages is a story about a struggle, a struggle whose requirements can exhaust and overwhelm participants. That people keep traveling through the labyrinth of cooperation—from harmony to conflict and back again, from hierarchy to equality and back again, from boredom to excitement and back again—is remarkable and seems to require explanation. What is it that keeps people participating in this amorphous, vexing, and sometimes threatening project? Why not just give up altogether, leaving behind the arguments, the compromises, the enervating details? Why not just "stay home"?

In the pragmatics of cooperation, the questions that participants articulate over and over emphasize the "how" of the process; these are questions that seek solutions: How can we achieve cooperation in this context of a wide-ranging heterogeneity of interests? How do we get all these differences (of people, institutions, nationalities, disciplines, components) to work together? How do we tame/unleash diversity in the interests of mutual production? In my time at ESTEC, I watched people labor to answer these questions in part by laboring to produce the artifacts they desired. Indeed, the artifacts became the answers. These artifact-solutions reminded and reassured participants that they could, in fact, work together, and so allowed them to begin again.

In this process, participants certainly had an interest in successful production, since it validated their existential involvement in the negotiations of working together. But participants did not continue to endure the oppression of uncontrollable structures, did not continue to argue, resist, join together, and compromise, simply to be able to go on "playing the game" of science (as Latour and Woolgar [1979] argue). This was not, or not only, a battle of interests and wills, where each scientist or engineer was out to best his or her opponent, marshaling support in an effort to defeat others (Latour 1987). The existential demands of productivity were often expressed in an instrumental language, such that all participants intended to "get something out of it"—whether technological expertise, data and published papers, or the maintenance of political and economic organizations and their positions within these—but this was not the only discourse. In listening carefully to the voices of participants, I became aware of another discourse, one intimating that the attainment of instrumental goals was not, in itself, enough to account for people's continually returning themselves to the scenes of struggle. Instead, participants' discourse suggested that they repeatedly involved themselves in the practice of cooperation because through it they were striving not simply to make something but to find something.

In working together, participants were dreaming about finding something other than space satellites, other than a unified Europe or even a functioning organization at the end of their travails. Cooperation indeed appeared to participants not only as an achievement but as an aspiration. The practice of working together was in this way not a mere technical exercise—a matter of connecting wires, organizing meetings, and launching satellites; it was, fundamentally, a sacred journey.

This sacred dimension of cooperation lies buried within the noisy and urgent practices related to the production of social order and the production of artifacts. It is neither irrelevant to it nor situated exclusively in the domain of structure. It is, instead, implicit in the whole system, in the cycles of order and disorder, in the very dialectic of production. Thus, although participants did not explicitly acknowledge the sacred aspects of cooperation in the process of working together, its presence was immanent in their confusions and their passions. It revealed itself most clearly in their ambivalent feelings about cooperation, which was where this ethnography began.

In this Epilogue, I return to the problematic of cooperation to ask, again, about the contradictions expressed by participants when they talked about something "out there" called cooperation. I suggest here that participants articulated an ambivalence about cooperation not simply because they participated in its production and so recognized its contingent quality (although this too plays a role); ambivalence stemmed as much from the difficulty they had in reconciling their understanding of cooperation as structure—as constraint, tool, and resource—with their experience of it as a kind of value. In this sense, the problematic of cooperation for participants concerns issues of transcendence and *communitas*, and the motivating powers of such experiences.

I am arguing that as people participate in the social system of working together, and engage themselves in the apparently mundane work of bureaucracy and technoscience, they are also articulating existential questions about unattainable desires and a vision of the good. This ideal is transcendent; for participants, it lies outside of or apart from social actors. Participants orient their practices according to this transcendent object precisely because it lies outside: its legitimacy derives from its externality. Simultaneously, this object reenters their practices as an underlying and unrecognized value. In this way, participants are at once desirous of this ideal and dependent on its legitimating capacity. In my exploration of this link between legitimacy (authority) and aspiration (to an ideal), I take off from the foregoing ethnographic analysis to pursue what I might call a "theology" of cooperation, an interpretive meditation on sacred striving.[1]

A Brief Digression on the Significance of the Sacred

Before delving into a discussion of legitimacy and aspiration in this particular social world, I want to address briefly the intellectual sources of my own thinking about the very idea of a theology of cooperation. This discussion owes much, once again, to Durkheim. Perhaps this is no surprise, given that in some respects, in this book I have traveled the same path that Durkheim took, as in his own work he was led from a consideration of the division of labor (in *The Division of Labor*, first published in 1893) to a consideration of the sacred in human life (in *The Elementary Forms of Religious Life*, first published in 1915).

For Durkheim, both the division of labor and religion held keys to understanding the connections between social and moral order in human society. In each case, Durkheim was preoccupied not only with the shape and structure of human connections, albeit in different domains, but with the way in which these connections constituted a fundamental source of morality, even an orientation to the transcendent. This morality could be understood, in part, as a domain of aspiration, as something ideal if not practically realized: "above the real world where . . . [a person's] profane life passes he has placed another which, in one sense, does not exist except in thought, but to which he attributes a higher sort of dignity than to the first. Thus . . . it is an ideal world" (Durkheim 1915: 469–70). Furthermore, the aspiration to this ideal was, for Durkheim, one trait that defined us as human: "The faculty of idealizing has nothing mysterious about it. It is not a sort of luxury which a man could get along without, but a condition of his very existence. He could not be a social being, that is to say, he could not be a . . . [human being], if he had not acquired it" (ibid.: 471).

The sacred, then, forms a part of social life generally speaking, and not simply of the religious life. It relates to something "which stands to us in the relation of moral ascendancy" (ibid.: 417). This standing apart, this separation, not only gives it the quality of an ideal, in Durkheim's terms, but also suggests its legitimating aspect, even though Durkheim's vision of the sacred omitted such considerations of power and politics.

To think about the problems of power and authority and their relation to the sacred, I turn to Durkheim's contemporary, Max Weber, who was, by contrast, preoccupied with precisely such questions. Although Weber did not discuss the sacred per se in his writings, he too, as Talcott Parsons ([1937] 1949) points out, developed a concept of the ineffable in human life. Instead of emphasizing the link between the ineffable and the ideal, however, Weber connected the former to the problem of legitimacy through the idea of "charisma." According to Parsons, this is "a quality of things and persons by virtue of which they are specifically set apart from the ordinary, the everyday, the routine" (ibid.: 662). Thus charisma, like Durkheim's sacred, was "set apart"; it is in this sense that charisma had an ineffable aspect such that it was "a quality, not necessarily only of persons, but of non-empirical aspects of the situation, of the action of, in a special technical sense, a 'supernatural' order, recognition of which underlies the moral legitimacy of normative rules generally. The concept, that is, becomes exactly equivalent to Durkheim's 'sacred'" (Parsons 1947: 75). The theology of cooperation that I articulate here makes this same link between aspiration and legitimacy.

Glimpses of the Sacred: Technology and Cooperation Revisited

In working together, participants have not only goals but dreams. They dream of transcending everyday worries, idiosyncrasies, and subjectivities, and in so doing, of being enlightened by nature's truths. For them, the transcendent is coterminous with an impersonal nature that they perceive to lie out there, always just beyond reach. This nature is the source of (universal) truth, and participants objectify it in their practices, conferring on it the power to affect their own actions since, as they see it, it lies outside their practices. Such a view was evident, for instance, in the words of an SSD scientist as he described to a scientific audience the design of a new instrument: "Nature has not been very kind to us, because there are few materials which can be used to observe in the ultraviolet." Here, the impersonal truth of nature stood between participants and their desires for understanding.

Nature's truths, moreover, could be strange and difficult, as well as exciting: "In research," explained an academic scientist, "you get unexpected results and you can't change that, even with all your knowledge you can't change that." For this reason, participants actively distinguished the objective and permanent "realities" of nature from the artificial and contingent subjectivities of the temporal, human world, a distinction made each time participants identified something as "real." This distinction was reflected in the way that SSD scientists routinely demarcated an objective understanding of nature, as represented in "scientific desire," from the subjective, those "inexplicable political factors" which interfered with that desire. During a seminar about research in a new field of inquiry, an academic scientist visiting ESTEC drew the distinction explicitly, arguing that the problem facing researchers in this area was "a human problem rather than a real one."

In participants' understanding of a transcendent, impersonal nature, human beings are excluded, as this quotation suggests. Nature, by contrast, appears to be the paramount domain of reason, pure and uncontaminated by human passions, interests, and needs (Zabusky n.d.). An ironic consequence of drawing this cultural dichotomy is that human beings appear to be "unnatural," since they and "their" world (a world that includes politics, identity, and emotion, among other things) are distinguished from a "natural" one. From this opposition, moreover, flows a series of cultural oppositions, all of which operate in the construction of the sacred that I am explicating here:

nature : human beings
: : rational : irrational
: : objective : subjective
: : real : artificial
: : significant : trivial
: : pure : contaminated

This concatenation of meanings suggests that human beings, along with the things they make and the things that make them, are impediments to enlightenment.[2] Or so it appears, at times, to participants.

Participants in working together strive to make it possible to discover the objective truths that a transcendent nature seems to hold. Nonetheless, nature itself does not appear on the level of working together. There, participants see instead the artifacts of technology and cooperation. These artifacts are like nature, however, in that they appear to lie outside the social process of working together, as I showed in chapter 7. This contributes to their power to legitimate local action and decisions—by appearing as "exogenous powers" (Atkinson 1984: 66), they are not implicated in the messy negotiations on the ground and so appear as impartial arbiters. But as I have suggested here, their power derives not just from their "charisma." The key to their local power is that they serve as the conduits to (transcendent) nature. In this way, the artifacts of technology and cooperation offer participants glimpses of the sacred in their everyday lives. In other words, the legitimating power of these artifacts comes not just from their constitution as structures lying "outside" the social process; it comes specifically from their contact with nature. As I indicated in chapter 7, these artifacts represent the achievements of social unity, of successful practice. But they also offer a promise beyond that of social connection—that of enlightenment.

In the following sections, I take each of these artifacts in turn, exploring how it is that they can promise the attainment of sacred dreams even as they also, in the guise of structures antithetical to practice, incite resistance.

Technology

In space science missions, technology is conceived explicitly to be a bridge between the questions of human beings and the truths of nature. The pragmatic goals of material production are in this way imbued with sacred meaning. In working together, participants' overwhelming interest in and anxiety about the details of the design and manufacture of technology reflect their recognition of this duality: technology as source of both instrumental return and sacred striving. Participants often talked about the way that technological artifacts could fulfill the promise of instrumental return; technology appeared in such moments to be the means by which participants could pursue a variety of (profane) interests, including securing promotions, positions, and prestige from its successful deployment. Yet participants also saw in these artifacts the promise of the sacred; in these moments, technology offered participants the possibility of escape from the confinements, contaminants, and confusions of the human world, escape into the realm of what they perceived to be the pure objectivity of nature. In this sense, it is technology that was "real" ("Now we're back to reality, describing hardware and software"), as measured against, for instance, the political interests involved in mission planning or the subjective "philosophy" articulated by mission designers.

Technology may be a mediator between human beings and nature, but it comes with a dangerous pedigree, made as it is by human hands, even though participants attempt to forget this in their practice. Its duality—simultaneously instrumental and sacred—threatens as much to prevent transcendence as to permit it. Indeed, the possibility that the technology-artifacts will obscure precisely what they are intended to reveal makes them objects not only of desire (participants described how they were "in love with" certain things, from ideas to instruments) but of anxiety. In order to prevent such obstruction, and in order to facilitate the enlightenment to which participants aspire, people work together not just to make technology but to make it disappear. Satellites, instruments, wires, chips—all of these must become transparent after they are made.

Scientists, for whom these artifacts are explicitly being constructed, want to be able to see through the technology they use, quite literally in the case of telescopes, and they manufacture designs with this transparency in mind. Engineers, for their part, strive to bring about a correlative invisibility by insuring the flawless launch of a satellite, which brings about the physical disappearance of technology, as well as the flawless operation of the satellite, so that scientists can think about other things. Those left behind, those anxious, expectant scientists, in turn, must repeatedly make the technology disappear conceptually if nature is truly to be observed. From their perspective, the more that they understand the technology—the more thoroughly they know its shape, principles of operation, and material limitations—the better they can ignore its effects, can account for its noise and so discount it in their efforts to discover nature's truths.

A good example of this drive toward transparency can be found in the technique of "calibration," which is a significant part of scientists' work in mission development. Scientists were continually striving to factor out the "noise" in the system, to distinguish, in the electromagnetic signals that reach them, what was an effect of technology, and what was "really" nature (see also Traweek 1988). This work preoccupied them throughout the process, from the moment of design, to manufacture, and even after launch. One SSD scientist I knew spent much of his research time calibrating signals from a satellite that had long ago dropped into the ocean. This persistent effort to remove all traces of a technological artifact from the "discovery" of nature, even though the artifact had long since physically disappeared, indicates the degree to which a concern with technology's transparency informs the technical practice of scientific work even after a mission is over.

The material artifacts that emerge from working together are both troubling and necessary. They are troubling because they bear the mark of human involvement in them (they are undeniably *made*) and because their very physicality threatens to impede participants' efforts to transcend the constraints of the human world. Yet at the same time, they are the only, or the best, means that people have to reach the impersonal perfection of nature.

Cooperation

Participants have an even more ambivalent attitude toward cooperation than toward technology. Unlike technology, cooperation is often viewed as fundamentally illegitimate. It does not simply get in the way of aspiration but is inimical to it, in part because the structure of cooperation brings stasis and hence death to the liveliness of working together. It appears, in other words, as the opposite and the enemy of the principle of action (conflict) that provides participants with the social and cultural means to escape from domination.

But those invocations of cooperation that appear in the climactic moments of decision making, when participants call on "history" or "the same boat" to legitimate their claims, articulate a vision of cooperation that is distinct from this understanding of it as a kind of structuring, constraining artifact. In invoking cooperation, participants are not simply making a claim, in a bid for authority, that cooperation exists; they seem also to be giving voice to an ideal of human connection, one that emphasizes participants' equality and connection in the presence of the truths of nature. They expressed their understanding of this ideal when they used terms like "community" and "family" to describe the social groups instantiated in working together, and when they insisted that values of "informality" and "spontaneity" characterized (or "should" characterize) their social relations. Thus, despite the recognition, even the celebration, of conflict and the noisiness of practice, participants have not thereby abdicated their dreams. Even as they revel in the chaos of enunciated differences, they are always striving to find a space, a social and a cultural space, in which they can make such noise recede or even disappear and so enact a free community of equals.[3]

This was made clear in a story told me by a group of academic scientists musing on the ideal of cooperation. The story concerned sixteen astronomers traveling together across Italy to observe a solar eclipse. The astronomers visited several cities along the way; during one of these visits, a fierce argument developed over how to spend the day, visiting churches or relaxing in cafés. Relations among the astronomers so deteriorated after this that by the end of the trip, no one was speaking to anyone else. On the day of the eclipse, the members of the group were scattered all over town, isolated and silent. Then the eclipse began, the darkness moving in, the birds' singing silenced. In the presence of this "impressive" event, the astronomers sought each other out, and everyone became "friendly" again, talking about what they had seen. The storyteller concluded that "something cosmological" could always bring people together.

This story reveals how participants imagine nature as a site in which human differences (of interest, needs, and emotions) can be transcended. It is in this way that they express their desire to work, even live, in a domain characterized by sacralized human connections and not simply instrumental ones. What I am arguing, in short, is that in the practice of working together, participants imagine *two* forms of cooperation, one pragmatic and the other sacred. At times these "cooperations" are conflated and appear to be the same—indeed, it is their conflation which leads to the manifest ambivalence and the apparent contradictions that people repeatedly express when talking about or alluding to cooperation. In effect, however, these two cooperations have distinct configurations and meanings.

The cooperation that has been the focus of the preceding chapters is "pragmatic cooperation." As I have shown, this cooperation is embodied in and represented by the bureaucracy of ESA and of the other organizations involved (including nations, industry, and universities), as well as by the diverse teams (e.g., the Project Team, the Science Team) and groups (e.g., the Ground Systems Working Group). This cooperation can be seen only retrospectively—we achieved that, hence we cooperated. People need these achievements, but they do not valorize them. Nonetheless, in working together, this pragmatic form of cooperation is often held up as a legitimate explanatory principle, in part because it provides a means by which people can pursue their different interests. Participants talk about pragmatic cooperation as being more "efficient" than competition or conflict; they talk about the way in which it insures the elimination of waste and duplicated effort. They talk, in essence, about its instrumentality. Participants submit to this cooperation because it is useful to do so; they choose to participate in this cooperation not because they value its silent order, but because they see in it a means for achieving their goals. In this sense, this cooperation is pragmatic.

Distinct from this kind of cooperation is what I call "sacred cooperation." This is a dream of cooperation as action rather than achievement; as such, this cooperation is always moving, always in potential and with potential. In other words, it is something participants are striving to achieve when working together; it represents not instrumental interests or practical concerns, but aspiration. Participants discover and rediscover this sacred domain in the process of working together; it is in the doing of the most ordinary tasks of technoscience, in the midst of what they thought was only instrumental and technical, that they encounter the sacred.

When participants talk about sacred cooperation, they talk about the way people connect with each other "naturally" and spontaneously: "You cannot put scientists in little cages," lamented an SSD scientist, because "science is a free thing." They argue that this cooperation occurs because of the way nature is; in this view, it is not nations, not organizations, but nature that requires people to come together in a certain way (in an egalitarian, informal way) to accomplish certain ends.[4] This is why an SSD scientist insisted, in the context of developing mission parameters, that "if scientific interest is the reason we come together, then we don't need rules," rejecting the bureaucracy and the treaties of politicians. Another scientist echoed these thoughts when he argued, "It's not Europe, but a collection of people coming together to solve a problem," a technical, rational problem relevant to the pursuit of nature's truths and not, or not only, to political ambitions.

Sacred cooperation is, in these ways, the opposite of pragmatic cooperation: it is liberating rather than confining, motivating rather than prescriptive, pure rather than polluted. The primary difference between these two modalities of cooperation lies in the cultural distinction participants make between social connections of two kinds: those constructed in terms of differences emanating from the political and institutional world (pragmatic cooperation), and those established in an orientation to transcendent reason (sacred cooperation). Put another way, pragmatic cooperation is the province of human beings; sacred cooperation is part of an impersonal nature.

Pragmatic cooperation approximates sacred cooperation because it is about making connections among human partners, because it suggests a social unity. But in the view of participants, this is an artificial unity; because it is made by human beings, participants regard it with suspicion, worrying about social hierarchies and the imposition and domination these bring. Pragmatic cooperation, then, does not represent the free and easy relationships that participants desire; such relationships can be found only in sacred cooperation, in which participants imagine human connections made without "rules," without nations, without hierarchies of control.[5] In sacred cooperation, people see themselves as coming together in a social intimacy made possible by their ongoing involvement in truth seeking. Sacred cooperation is, thus, an aspiration—it is about connecting (potential), rather than connection (achievement); in this way it is, for participants, the path to both communitas and truth.[6]

The Path to the Sacred

The sacred domain resonates in the shadows of the artifacts that people have produced, tantalizing them, drawing them onward, but ever receding. To get there, to tread the path to the sacred, requires working in the world. But for these participants, it is not just any kind of work that promises to bring them closer to nature; it is only what they call "real work" that will help them to discover what they consider to be the objective truths of nature, and what I might call enlightenment of a more sacred kind. Engagement with real work is, therefore, necessary in the striving for the sacred. It is also difficult—it is difficult to find the kind of space, not to mention the kind of self, conducive to real work. Finding it requires constant vigilance combined with the faith that such a space can be inhabited, however briefly, in the human contexts of working together.

Real Work

Participants talked about "real work" all the time. What is real in this case is not opposed to something illusory, but instead to something superficial, even artificial; as I noted above, they contrast the "realities" of nature with the artificiality of the human world. What is real is what is significant and essential, what really matters, where the substance is. In opposition to this were other kinds of work that participants considered to be "political" or "professional," work that entailed all manner of details in themselves basically "trivial." Real work was work engaged in without unnecessary distractions from the profane world. It was "the real thing that can keep you alive"; it was, as one scientist said of his data, "like oxygen"—accordingly, participants give themselves up to it spontaneously, breathing in it and through it. Real work is, in this way, the vocation: it is what calls people to that kind of practice (Weber 1958). As such, real work is the path to the sacred.[7]

SSD scientists in particular talked about "real work," or else about "really getting down to work." For them, such phrases referred to hands-on activities dealing with science and technology, not to tasks involved in what they, usually derisively, termed "science support," "scientific management" or "administration," "politics," or even "engineering." Real work could involve research tasks such as collecting and analyzing data, or any activities relating to the design or development of instrumentation when the technical details were paramount. There was undeniably a social component to this type of work, in that it involved collaborating and interacting with colleagues, but when participants identified something as "real work," they meant that the focus was on the data and not on the collaboration, or that the questions were about technique, requirements, specifications, but not about relationships.

What defines real work is this apparently impersonal and "pure" quality. Real work eliminates the noise that comes from the effects of human involvement in the world, an involvement manifested in the subjectivities of personal passions, national identities, and social hierarchies.[8] In real work, such distinctions are set aside as participants try to focus on the work at hand. Real work thus appears to be uncomplicated by anything but technical details, whether these be details of mathematics or technology. It is unaffected by the noise that surrounds it on all sides, noise emanating in part from human differences that, in the "objective" realm of nature, are at best irrelevant and at worst contaminating.

All the other kinds of work available to participants are, by comparison, contaminated versions of work, precisely because they must take into account and focus on the sources and elements of noise. For instance, professional work that takes into account not only the interactions and decisions of nations but the interactions and decisions of agencies, universities, and industrial organizations is characterized by participants as "political"; the political is so much noise when it comes to real work. This does not mean that participants ignore or discount such factors; as I have shown in the preceding chapters, participants are cognizant of these elements of their work and indeed make use of them, deploying them strategically as resources in the assertion of difference. In real work, however, people focus their efforts on eradicating these noisy elements.

All participants laid equal claim to the importance of their dedication to "real work." When scientists and engineers opposed each other, for instance, they did so in part by denying that the other understood what constituted this real work. Scientists argued that they strove to keep their minds free from "the nuts and bolts" that "cluttered" the minds of engineers. They preferred to work without the "extraneous" constraints imposed by politics, finances, and even technological components when these were viewed as resources deployed by engineers in a political manner. Engineers, by contrast, treated such "nuts and bolts" as the essence of real work. For them, engineering work began with and took into account these constraints; learning to deal with and incorporate these in the design and manufacture of technology *was* real work. By contrast, they rejected the confinements of paperwork that enforced their separation from technical details as they monitored contracts instead of doing what they considered to be "real" engineering.[9]

The point here is that although people did not agree on what constituted the content of real work, they did agree that real work was the only kind of work worth doing. What they seemed to be saying, in their insistence on the importance of real work, was that it was the only kind of work that was significant, pure, and free from noise, however that noise was substantively defined. Other kinds of work were most relevant to the domain of pragmatic cooperation, where they were required by those "artificial" constructions such as nation-states and bureaucracies (which were not, in this view, "real," that is, found in nature). Participants acknowledged that they must undertake this other work in order to be able to do "real work," but other work was not, therefore, valued: "The name of the game is industrial—this doesn't mean you can't do science in this context." Indeed, these other kinds of work were dangerous, because they threatened to suffocate people by limiting their ability to breathe in the "oxygen" of real work. For this reason, other work does not offer the enlightenment that participants are seeking in the substantive truths of real work.

Vigilance

Engagement with real work brings exacting requirements, demanding of participants a constant vigilance in the effort to eliminate noise from the domain of real work. Everything, for instance, must be kept clean. Cleanliness was demanded most particularly of the material objects of technology, much of which could be manufactured and tested only in "clean rooms." Clean rooms controlled the atmosphere of the physical environment and insured that dust and dirt remained outside so as not to interfere with the operation of delicate components. Participants donned special coats and shoes to enter these rooms, purifying themselves before coming into contact with technological components that in turn come into contact with nature.

This rational act was not simply a matter of secular technique; it was a sacred duty. As I see it, such cleanliness is demanded of human beings in their efforts to approach transcendent truths. Accordingly, participants' commitment to real work requires them to extend their purifying efforts to their human selves. It requires that people strip themselves down as much as possible, eliminating worldly preoccupations and distractions from their physical spaces, their language, and even their social selves.

In the work space of ESTEC, for instance, the long, monotonous corridors and dull colors of the original buildings appeared to many scientists and engineers as the only appropriate environment for real work. By contrast, the playful design of ESTEC's new office towers dismayed some participants. These buildings did not fit the mold of the old office complex, with their soft wood tones, odd angles, and staircases that terminated unpredictably in the middle of the building. All of this was simply too distracting. One scientist insisted that he actively preferred the predictable staircases and orderly right angles of the original structure; for him, these signified not monotony but a purity—specifically, a purity of function—that played a critical role in helping participants focus on their work.

Participants stripped down their language, too, by attempting to focus only on communication, as if language were there only to permit the unambiguous exchange of information. The proliferation of acronyms, in speech as in documents, represented most starkly this stripping down, when words themselves were eliminated in an effort to remove the messy potential for connotation instead of denotation.[10] The emphasis on jargon and on the use of technical terms also admitted no emotion or passion into talk about work. Even the common use of English betrayed a concern with this kind of straightforwardness, as people stressed the communicative function of language to the exclusion of other aspects. Everyone talked in a stripped-down fashion, suppressing the use of subtle nuances, metaphor, slang, or idiomatic expression. Even native speakers of English purged their speech of color. People spoke quickly and in technical detail, but they did not wax poetic, even in defense of interests. When they argued, they made their points bluntly. In other words, participants preferred, in working together, a language that was direct and simple, and they tried in many ways to make their language as pure as possible in their efforts to focus on real work.

In the expression of people's social selves, finally, this stripping down was most evident in the denial of identities and interests deemed irrelevant to real work. Nationality, for instance, was repeatedly pushed aside in work contexts. Although its appearance in the canteen or during coffee breaks might be acceptable (as I showed in chapter 6), participants insisted that when it came to "getting down to work," differences in accent, dress, or body language were only insignificant and superficial elements of style. People endeavored to make their national identities, and whatever passions or interests might stem from these, irrelevant.

Even occupational identities were ideally to be purged from the spaces of real work. It was only by inhabiting these identities that participants defined social contexts for work, yet when it came to "really getting down to work," they tried to ignore professional categorizations such as "scientist" or "engineer." They preferred instead to emphasize the common problems that lay before them, solutions to which depended on embodied but disinterested expertise. Indeed, people worked to liberate expertise from those institutional contexts that inhibited its true fulfillment. When it came to engagement with real work, participants recognized that people had to divest themselves of these identities of interest—whether national or occupational, such identities were viewed as noise that interfered with the work at hand—because these were not given in nature but stemmed instead from the artificial world of pragmatic cooperation.[11]

In their repeated efforts to strip themselves down in these diverse ways, participants endeavor to approach closer and closer to a pure objectivity, since they believe that only in objectivity is the truth of nature revealed. It is a strenuous effort, because the artifacts of the human world repeatedly intrude and must be continually swept away. So it is that participants spend an inordinate amount of time and energy trying to clear out a clean and pure space—a physical, cultural, and psychic space—where they can come together unencumbered by the subjectivities inherent in national, disciplinary, professional, bureaucratic, and institutional affiliations and the passions attendant on these. They are striving to reach that space where they can be, instead, harmonious, spontaneous, and equal in their shared dedication to real work.[12]

Faith

Despite this vigilance, the achievement of "real work" is largely illusory; there are no "pure" scientists or "pure" engineers doing "real work" anywhere. As I have been at pains to demonstrate in the preceding chapters, the noise of external circumstances and defining conditions is ever present, never entirely eradicated from some inner sanctum in which the messiness of the personal and the political disappears. Participants themselves realized this. Indeed, as I have shown throughout, it was almost impossible to talk to them about their work without hearing about the constraints, impingements, and influences emanating from professional and political structures. It was clear, moreover, that participants derived unmistakable pleasures from their involvement in and negotiation of these "profane" elements of working together, in particular, of the differences that they so often embraced and asserted to one another (as I argued in chapter 6). There was momentum, even excitement in the disorderly unfolding of practice, an excitement that signified the presence and power of diversity. Participants asserted diversity as a moral stance in opposition to the oppressive unity of structures of cooperation, and they valued it as a principle of action that required them to immerse themselves repeatedly in difference. It was, indeed, their differences that helped participants remember that they were alive.

In other words, participants did not reject the complications that constituted working together, nor did they have any illusions about the power and importance of these complications. What they did have was a kind of faith that they might transcend these human "trivialities" in a space they called real work. The persistence of this dream of purity coexists in the practices of working together with the articulation of diversity as a key value. The presence of the one value does not contradict the other or render it irrelevant. They are not contradictory so much as oriented toward different domains in the unfolding temporal arena of practice. People are committed to both, not simultaneously, but through time.

When I say participants had faith, I am arguing in essence that in and around talk about real work, about purity and objectivity, what emerges most eloquently is desire, a longing for a space in which "real work" can take place. This longing is connected less to the excitements inherent in pragmatic cooperation and the push to production than to the idea of purity that participants understand to be integral to sacred cooperation and the enlightenment it promises. This is why participants repeatedly try to strip themselves down, to divest themselves of diversity, as it were; they do this in order to be able to enter into this sacred domain of cooperation, where it is not their differences that matter so much as it is their fellowship.

In talking about real work, then, participants are not (always) making a legitimating claim that they have actually succeeded in finding the time or space to do it; they are, instead, articulating a regret, a regret that, ultimately, they are forever excluded from the pure space of sacred cooperation. They are excluded, culturally speaking, because when they express their passion for objectivity, they give voice to a kind of human desire, a personal emotion that, in this cultural worldview, is incompatible with, or at least distinct from, the impersonal ontology of a transcendent nature. The longing for real work reveals a paradox: it is participants' fundamental humanness that makes them unable to enter into the transcendent realm of nature. Real work can be, therefore, only a dream of sacred cooperation, in which egalitarian and spontaneous social relations are achieved effortlessly, insuring the discovery of transcendent truths.[13]

By identifying "real work" as a utopian space, I am suggesting that what motivates participants to keep working together is sacred aspiration—and the corollary regret that the sacred remains beyond reach. Participants are not satisfied that they have attained enlightenment or even the purity it demands; nonetheless, they remain obsessed by the *hope* of finding it. It is thus in these invocations of "real work," and the corresponding dream of purity, that a theology of cooperation can best be discerned. Purity is sacred in two respects: it is a source of legitimacy, suggesting the power of impersonal nature to dictate human practice; at the same time, it is an object of desire, as people dream of making the right kind of human connections, which will permit them access to transcendent truths. Such dreams are a site of struggle, since the very things that people value in one domain seem to impede their connection with another. All that people *can* find are glimpses of the sacred, barely discernible, a mirage of freedom shimmering around the contours of more oppressive shapes.

In the mundane contexts of working together, people are always striving to find that utopian space of easy communion. As a space of and for dreams, sacred cooperation is an aspiration, and so materially unattainable, which is as it should be. Sacred cooperation is not, in other words, the same thing as the enduring state of being enabled by and enshrined in concrete structures; these are indeed the pitfalls of human efforts at constructing relationships of cooperation. Whereas these structures are achievements of unity, sacred cooperation is the potential for union. Sacred cooperation is motivating precisely because it therefore demands constant striving for something that is (almost) always beyond reach; in this way, it gives direction, movement, and meaning to life and work.

Participants keep working together because it is only in the doing, in practice, that they have even a chance of experiencing such blessed moments of connection, when the sacred erupts briefly into the profane world of work. In such moments of intimate and collective concentration—which do, in fact, occur, however ephemerally—participants experience the epiphany of creativity and touch for an instant that breathtaking purity they so desire. Such glimpses and experiences sustain and motivate them as they travel through the labyrinth of cooperation. Participants keep working together, keep enduring structures of cooperation and risking the dark side of diversity, because they are always trying to find that sacred space where what is inessential falls away and they are able to be together before nature.

Cultural Resonances: The Sacred Dreams of Modernity

In the theology of cooperation I have explored above, legitimacy and aspiration both depend on an idea of purity—specifically, the purity of the "real." What is real does not vary; it endures, unaffected by external influences, particularly those emanating from human beings. This idea of purity is not peculiar to technoscience, even if the version I have explicated here focuses on the dreams of its practitioners. Rather, it is a key component in the "ideological arena" in which European space science missions are formed. Hess (1991: 5) defines such an arena as the "confluence of social and cultural systems" in a historically and culturally situated space. In this case, that "space" is Europe in the late twentieth century, a space paradigmatic of modernity. ESA emerges here as an especially triumphant production and purveyor of this paradigm. This is because it represents a nexus of three of modernity's defining institutions—nation-state, bureaucracy, and technoscience—which come together in ESA in a mutually implicating, reinforcing, and codetermining relationship. ESA is so successful at what it does, indeed, because this nexus sets off a profound cultural resonance, one that vibrates around a shared idea of purity.

In the preceding pages of the Epilogue, I have focused attention particularly on the way in which scientists (and their engineer colleagues, to some degree) imagine science as a sacred form of activity, and a sacred form of relationships, leading to enlightenment. In this imagining, science is and must be protected from the contamination inherent in everyday life. This protecting is undertaken not only by scientists and engineers; it is a shared mission of many observers of science as well, including, as Bloor (1991) and Fuchs (1992) have argued, philosophers and sociologists of science. The ideology of the pure objectivity of science is well-rehearsed in the literature on the social studies of science (see, e.g., Restivo 1988). Here, I turn my attention briefly to the way in which both nation and bureaucracy, too, acquire the status of sacred object, in order to make audible the resonances across these domains.

In the ideology of science, it is nature that endures; in the ideology of nationalism, it is the nation. It may seem paradoxical, at first, that what from the point of view of science appears as a source of contamination—the nation as contingent phenomenon, a product of irrational human beings rather than of impersonal nature—should appear in nationalism as sacred because it is essential and noncontingent. But one system's artificiality is another's authenticity. Or at least, as Herzfeld (1987: 155) wryly observes, "whatever the actual state of affairs . . . official discourse has certainly succeeded in making nationality [and nations] *look* absolute" (emphasis in original). The nation is more than simply "official discourse," however. As Benedict Anderson (1991) has so persuasively written, the nation is an "imagined community" as much as, if not more than, the site of political strategizing. As such, it is a source of meaning. It promises, in particular, the unity of "deep, horizontal comradeship" (ibid.: 7); it is this feeling of authentic connection "that makes it possible . . . for so many millions of people, not so much to kill, as willingly to die for such limited imaginings" (ibid.).

The power that the nation has to inspire such passions of commitment comes from its "transcendent status," as Herzfeld (1992: 6) has argued. The nation is set apart from the ordinary, even, as Kapferer (1988: 1) indicates, "plac[ed] . . . above politics" (ibid.: 1), despite the fact that nations and states (that quintessential domain of politics) invariably and unavoidably go together. Politics is the domain of division and difference; it reflects the corresponding contingencies of interests and influence in ongoing competitions for control. The imagining of a sacred nation emphasizes, by contrast, connection; it elides the differences that are so salient in "politics" in a promise of unity. In this way, the sacralization of the nation represents a kind of purification—specifically, the eradication (and denial) of people's differences. In the nation, then, we find another version of "the theology that equates actual social and historical experience with spiritual or cultural imperfection" (ibid.: 32); it is people, with their "flaws, social life, [and] sex" (ibid.: 46), with their humanness, who continually threaten to undermine the sacred connections offered in the communitas of the nation.[14]

The cultural systems of nation and science, thus, both resonate around the value of pure objectivity. Both establish themselves as fundamental forms of authentic experience and activity by constructing their essential naturalness against the artificial constraints and developments of human history and human interests. To insure this, neither nation nor science, in other words, can appear as artifacts (having been made); they must, instead, appear as objects (found in nature). Subjectivities, and their attendant differences, are therefore banished from the discourse that imagines these cultural domains. Thus, both of these cultural visions of objectivity, albeit in different ways, demand the purging of contextual (that is, human) contaminants, one to insure unity, the other to insure truth.

The cultural system of bureaucracy operates in much the same way; here, too, human beings are problematic. People, with their interests, their desires, their idiosyncratic needs and imperfections, confuse the rational order bureaucracy promises to provide. For this reason, as Weber (1946: 215–16) indicates, the "specific nature . . . [of bureaucracy] develops the more perfectly the more the bureaucracy is 'dehumanized,' the more completely it succeeds in eliminating from official business love, hatred, and all purely personal, irrational, and emotional elements which escape calculation." In its separation from, even disdain for, human beings, bureaucracy also claims for itself transcendent status. Herzfeld (1992) explicitly argues, in fact, that bureaucracy has the power of the sacred; he even likens European state bureaucracies to "the ritual system of a religion" (ibid.: 10), specifically the religion of nationalism in this case. Bureaucracy, as nationalism, emphasizes "the separation of eternal truth from the mere contingencies of society and culture" (ibid.: 19).

Bureaucracy, then, is the instantiation of an effort to delimit a "pure" space for rational action and decision, abstract and universalizable, purged of the imperfections of human beings.[15] The effect of bureaucratic labors is to keep decisions clean, that is, free of the contaminating influences of human particularities. This is what Brenneis (1994: 31) calls "the social production of impartiality," in which emotion, idiosyncrasy, and the subjective differences of various identities and interests are all kept at bay. The resulting impartiality is, however, more than just a technical outcome of such labors; it is, as Brenneis points out, a key element of "democratic values of due process and equal treatment" (ibid.: 25). I would argue that this impartiality, like unity and truth, is best understood as another form of purity that must be protected from contamination; that is, the valorization of impartiality reveals bureaucracy's sacred promise.

What I have argued here is that modernity's dreams take the shape of purities that resist the onslaught of difference and subjectivity. This dream of the "real," one that, as Herzfeld argues (1987: 19), "places rationality above and beyond experience, transcending the particularities of historical time and cultural place," has a long history in Europe, going back even to fifth-century Greece. Its modern configuration took shape in the sixteenth and seventeenth centuries, when Reason gradually came to replace God in the sacred focus of the West (Dumont 1986). During this time, a "concordance" (Tambiah 1990: 12) developed between certain aspects of Protestantism and the "ethos" of capitalism (Weber 1958), a concordance seen also in the development of science (Berman 1981, Merton 1968, Tambiah 1990) and bureaucracy (Collins 1975). What linked these different sociocultural systems was a sacralization of rationality. In nineteenth-century Europe, this sacralization of rationality continued to spread, reaching its apotheosis in a variety of "totalizing ideologies." Chief among these was nationalism, which stressed "absolute knowledge . . . [that] assumes the utter irrelevance of the observer's own context" (Herzfeld 1987: 14). Such absolute knowledge was perceived to be the special property of rationalized institutions, such as science, nation-state, and bureaucracy, each of which claimed, culturally speaking, an authentic connection to the enduring objectivity of nature.

I am not here making an argument about causes but about a cultural logic; my intention is to point to what is shared across these different institutional domains. I am arguing, in other words, that an orientation to and desire for purity continue to circulate in the "ideological arena" of modernity, imparting sacred status to its institutions. Pure truth, pure unity, pure impartiality—all of these are naturalized and sacralized in modernity's institutions, thus legitimating these institutions' claims to transcend the vagaries of local knowledge, social difference, and personal experience. Such concerns of everyday life—those of interest, identity, and emotion—appear, in the pure space of reason, not as essential components of life, but only as "matter out of place" (Douglas 1966); it is rationality—the quintessence of modernity—that, apparently, has permitted the elimination of these concerns. Modernity's controlling power rests on this ideology of rationality, which describes rationality as a powerful technique, independent of emotion or other subjective interest, promoting efficiency and objectivity in the control of nature. The power of this rationality comes from its ability to allow people to bracket action out from context, and so to calculate within narrow, "technical" parameters. In this way, rationality appears as a kind of purification, one that constructs, as Eisen (1987: 112) argues, a "free space of action [that] . . . is deemed essential to efficient functioning."[16]

The freedom that this rationality promises is, however, constructed at the expense of all that is human and subjective (Habermas 1989b); in this way, the spread of rationality has been part of the modern transformation of the world from an active agent to a passive receptacle. This is, in part, what Weber meant when he talked about the "disenchantment of the world," recognizing in this transformation both possibilities and dangers. Nowadays, however, the possibilities seem barren; instead, as philosopher Charles Taylor (1989: 500) indicates, the idea of "disenchantment" articulates only a modern lament. Taylor hears this lament when people in industrial societies talk regretfully about their sense that the world has lost something; no longer "a locus of 'magic,' or the sacred," the world has come "simply to be seen as a neutral domain of potential means to our purposes." The neutralizing of the world is, in this view, a direct result of the profligate application of "disengaged, instrumental reason," the kind of rationality characteristic of, and perhaps necessary to survival in, capitalist and technological society. However necessary it may be, Taylor suggests, people perceive this "mode of existence" as empty because in it the world has lost not only its magic but the richness of symbolic meaning; all it has left is the domination that accompanies instrumental calculation.

This lament stimulates much contemporary scholarship. One trend is to take modernity to task for the way it misrepresents social relationships; scholars expose how the rhetoric of states, bureaucracies, and science, for instance, masks elite interests and relations of dominance. Such analyses argue that rationality is merely an ideology, leading not to some reality of reasonable or efficient action but only to continued oppression and/or the furthering of elite interests. Moreover, such analyses continue, the resulting policies, techniques, or actions are "actually" irrational in effect (e.g., Herzfeld 1987, Restivo 1988). Another trend is to critique the way those in power impose standards of instrumental rationality onto populations that might otherwise orient their practices according to some other mode of thinking. These analyses argue that such people resist this imposition by continually recognizing "enchantments" in the world around them, even if the space of enchantment is ever shrinking in the face of the exercise of rationalized power (e.g., Holmes 1989, Taussig 1980).

These critiques, as important as they are, depend for their power on an often unexamined corollary: that only "others"—whether European peasants, non-Western villagers, the colonized, the working class, the masses—find enchantments in the world. "We" (Western, modern, elites), on the other hand, supposedly see only a desiccated world, one defined by efficiency and instrumentality, abstraction and order. Apparently, "we" have no culture, only ideology, since we are the inhabitants of a world dominated by rationality. But this kind of view rests on a false dichotomy, and a pernicious one at that, as Jean Comaroff (1985) has argued. Indeed, it is this view—that the others lead enchanted lives—that in the past denied all these "others" the capacity to reason, leaving them with only "superstition" and "irrationality," and making them available for "rational" domination. Decades of ethnographies have shown, however, that "the natives," wherever they are found, have as much practical reason as we do, even if they also find enchantments in their world.

The "others" thus emerge from this scholarship of lament as complex people, whose lives are all the richer for the complex entanglements of reason and enchantment they contain, and for their efforts to resist the domination of modern, Western rationality. In so enriching these lives, scholars have tended to embrace the assumption I describe above: that those on the "other side" of the unspoken dichotomy (to the extent that this dichotomy is reified in analysis), those who live and work in the rationalized institutions of modernity, live unsparingly disenchanted lives, and perhaps even prefer it that way. Comaroff (ibid.: 144) urges us to recognize, however, "the fallacy of assuming that Western consciousness is itself depersonalized in any simple sense." Indeed, in her analysis of Tshidi consciousness, she argues that "no simple distinction between instrumental and symbolic practice makes sense here, *or indeed anywhere*; instrumental action is always simultaneously semantic, and vice versa" (ibid.: 125; italics added). What Comaroff seems to be saying is that although the ideology of rationality suggests that instrumental practice is distinct from and devoid of symbolic meaning, careful cultural analysis of actual practice counters this suggestion. The very idea of a pure instrumentality is itself symbolic; indeed, in our privileging of efficiency we articulate also a semantics of efficiency—rationality is meaning*ful* not meaning*less*, as Western laments about "disenchantment" would suggest. It is for this reason that Herzfeld (1992: 65), in his analysis of "rationalized" state bureaucracies, cautions us: "The fact that the modern world appears disenchanted in relation to its seemingly reverent forerunners should not lead us to take its rationalist claims too literally. It, too, rests on utopian and cosmological foundations."

For these reasons, I insist that despite the sense of loss "we" may feel, we must also recognize the presence of enchantments in lives lived in the modern age. The enchantments of modernity are, however, difficult to find, in part because they are not located in nature per se but emerge, as I have argued, in the ongoing negotiation of social and technical connections that constitutes daily practice in a wide variety of institutional domains. In other words, the enchantments of modernity are obscured in part because they emerge in practices that participants often claim (and observers believe) to take place, exclusively, in the domain of "rationality." Such claims represent, however, the *ideology* of rationality, the very idea of which denies magic in the world; they seem to deny that there can be any "culture" in the world of the modern West.

The ethnography of European cooperation in space science that I have undertaken in this book represents, however, my conviction that there *is* "culture" even in modernity's rationalized institutions, and I have endeavored to delineate the shape that meaning and enchantment take for those living and working at the heart of modernity—in this case, in the quintessentially "rational" world of ESA space science missions. Here, everything often looks exquisitely rational; when participants locate reason in nature and then desire connection to nature, they appear simply to want to be reasonable. This seems a quite practical affair, apparently stripped of fancy accoutrements and devoid of ritual encumbrances or mystifications. But such stripping down, as I argued above, is itself a sacred act. In this way, the exercise of rationality in working together represents not only legitimation, but also aspiration; it is a kind of sacred striving through which participants endeavor to attain that purity they so desire. Thus, modernity's enchantments emerge in the work of rationality itself. It is for this reason that cooperation—that which is at once sacred and pragmatic—must be interpreted through a theology that links real aspirations with the exercise of legitimation in modernity's dreams of purity.[17]

In working together on space science missions, participants who in so many ways can be identified as elites (as the keepers and purveyors of "formal" knowledge; see Foucault 1980) themselves experience chaotic uncertainties and perceive sacred meanings in the world, just as do the "others" of contemporary scholarship. They, too, want to distinguish themselves from the state and bureaucratic structures that threaten to pin them down and render them voiceless and immobile. In the practice of cooperation, participants keep themselves free by resisting one kind of unity, the unity of a cooperation that oppresses and homogenizes. But they simultaneously keep themselves free by expressing their own desires for union and by constantly striving for sacred connections—even if this striving is sometimes accompanied by a discourse of cynicism, since everyone knows this goal cannot be attained. We cannot dismiss such aspirations as epiphenomenal, mere ideological veils hiding instrumental goals of domination; as Taylor (1989: 519) argues, "not only can some potentially destructive ideals be directed to genuine goods; some of them undoubtedly are." In the ideals of truth, unity, and impartiality, these participants identify a world resonant with meaning; in their continual striving to reach these, they give voice to sacred aspirations, to dreams of communion and epiphanies of collective creativity in work.

Notes

Introduction: Multiple Cooperations

1. In the spring of 1994, the European Community changed its name to the European Union. During the time of my fieldwork, however, it was the idea of the "Community" that gave meaning to the practices and structures of cooperation I learned about at ESTEC. For reasons of ethnographic accuracy, then, I retain the designation *European Community* (and *EC*) throughout this text.

2. For instance, differentiating movements proliferated in the 1970s, in renewed ethnic and intranational regional movements for autonomy (e.g., Basque independence), suggesting that people wanted less, not more, unity; at the time, this made the movement toward European integration seem moribund (see, for instance, the articles in Esman 1977). Nonetheless, the EC continued to grow. In the 1980s, the resistance of Britain under Margaret Thatcher to join wholeheartedly in key areas of integration seemed again to suggest that the ideals of Robert Schumann and Jean Monnet (the architects, in the 1950s, of the EC) would remain unrealized. Yet during this time, the EC member states signed the Single Market Act, to take effect in 1993. In the 1990s, local populations have increasingly voiced their objections to the decrees emanating from Brussels (or Maastricht), and governments are pulling back from, among other things, the European Monetary System, cornerstone of integration. All of this seems to call into question the viability of integration, yet I would argue that these resistances can also be viewed, in a significant way, as themselves necessary parts of the integrating tendency (Gerlach and Radcliffe 1979).

I should also point out that alternatives to the nation-state version of European integration are being pursued today, although I do not deal with these in this book; see for instance Galtung 1989 and Stephens 1993 for discussions of a variety of "European projects," including regional and "green" or "peace" models of integration that cut across nation-state boundaries, dispensing with them in the process. Holmes (forthcoming), in a less positive vein, offers a critique of the very idea of European integration, arguing that it contains and constructs a radical, corporatist logic dangerous to difference and democracy.

3. A special issue of *Science* (vol. 237, no. 4819) in 1987 presented an overview of European scientific research and articles on each of these organizations. The trend toward big science and the concomitant need for integration of disparate participants is not peculiar to Europe; the United States has in the past been at the forefront of such massive undertakings, whether in the Fermi National Accelerator Laboratory (Fermilab), the space missions of NASA, the Tokamak Fusion Test Reactor, or the human genome project. For historical and sociological discussions of the growth of big science, see Price [1963] 1986 and Galison and Hevly (1992). Recent events in the United States suggest that the time of unlimited support of such technological and scientific superprojects may be on the wane; Congress canceled the Superconducting Super Collider

project in 1993, and NASA has been forced to curtail various projects and to seek international cooperation to make what projects remain affordable (e.g., both the space station and the Hubble Space Telescope are joint ventures, involving among others the European Space Agency).

4. This kind of attitude was made explicit during ESA's twenty-fifth-anniversary celebration, when a series of European leaders from government, industry, and science spoke precisely about the ability of science and technology to bring people and states together politically. In the Jubilee publication, then director general Reimar Lüst (1989) reflected on the technical and scientific projects completed by the Agency: "That all of this has been achieved through the united efforts of nations and cultures long-split by enmity and suspicion is sometimes forgotten, yet European space ventures have shown just how barriers can be lowered in pursuit of peaceful objectives. That political will, scientific curiosity, and industrial know-how can be harvested for the common good." I discuss these views in more depth in chapter 2. I should add that this attitude is not peculiar to Europe or to space science. In Zabusky (n.d.), I discuss how American scientists articulate a belief that the rational pursuit of scientific knowledge can serve as a model for the achievement of peace in the global political arena.

5. Research in the sociology of organizations examines, among other things, the way in which bureaucratic organizations have as many irrational as rational components; the way official organizational frameworks are inhabited by "informal" organizations; the way people manipulate and interpret organizational rules and structures; and finally the presence and production of organizational culture (see Morgan 1986 for one overview of approaches to organizational theory; for studies in "organizational culture," see the papers in Frost et al. 1985 and Pondy et al. 1983). From a radically different perspective, Certeau (1984) discusses the subversive "tactics" of those people who "consume" the structural orders imposed by technocratic institutions; for instance, he describes "*la perruque*," the efforts of workers to "divert time . . . from the factory for work that is free, creative, and precisely not directed towards profit" (ibid.: 25). Anthropologists have devoted less attention to the study of bureaucracy and organizations, perhaps reluctant to concede that there might be "meaning" in such an artificial and rationalized environment. In fact, as organizational sociologists have noted, "culture" abounds in and around bureaucratic organizations, although these sociologists tend to regard culture in a reified way. For more interpretive cultural analyses of bureaucracy by anthropologists, see Britan and Cohen 1980, Handelman and Leyton 1978, Handelman 1981, and Herzfeld 1992.

Chapter One
The Study of Cooperation: Theoretical Issues

1. At the same time, it may be no accident that this approach makes so much sense in relation to my ethnographic data; the anthropological debate about the nature of social structure itself is in part a legacy of Durkheimian sociology's emphasis on order and social integration (structural forms of "cooperation"). Both Giddens and Bourdieu certainly draw on such decades-long

debates (directly and indirectly) in their own theoretical elaborations of the problem of structure. The distinction they make between structure and practice, for instance, resonates with similar distinctions made in the past by other anthropologists who have drawn in diverse ways on Durkheimian principles. Even Radcliffe-Brown (1952: 4), quintessential structural-functionalist, noted that "the concrete reality with which the social anthropologist is concerned . . . is not any sort of entity but a process, the process of social life. . . . The process itself consists of an immense multitude of actions and interactions of human beings, acting as individuals or in combinations or groups." He distinguished such diversity from what he called "a form of social life," which corresponded by contrast to the "discoverable regularities" reflected in that diversity of everyday life; he depicted anthropology, however, less as an investigation of the "diversity" than as the "comparative theoretical study of forms of social life." French structuralism, too, considered an analogous distinction in its search for universal "deep structures" of mind that lay beneath the surface profusion of differences; Lévi-Strauss (1963a: 279), for instance, even argued that the abstractions of structural systems have "nothing to do with empirical reality but with models which are built up after it." Nonetheless, although anthropologists have long realized that "empirical reality," "everyday life," or "the facts" are redolent with indeterminacy, ambiguity, and contradiction, in the past many have tried to resolve these confusions of everyday life in various reductive scenarios of equilibrium (this was true, for instance, of the structural-functionalists), rather than trying to capture or explore the energy of the dialectic itself.

2. Since I imagine that this book is being read by scholars from various disciplines besides my own, I feel perhaps that it is necessary, given Traweek's (1992) experience, to signal that my invocation and citation of Durkheim here and elsewhere in this text represents my understanding of Durkheim as, in Traweek's (ibid.: 437) words, "a kindly gent, one whose name in our texts is . . . a reminder that we do not have to be either materialists or idealists." This means, too, that I borrow some of his insights without subscribing to the entirety or specificity of his argument. I do not intend, in other words, to advocate what some may view as Durkheim's "decidedly archaic . . . position about ideas and action," nor, therefore, "to defend that territory with him" (ibid.).

3. This is the case as well in other acephalous societies such as those discussed by the authors in the volume edited by Brenneis and Myers (1984). These authors describe a variety of societies in the Pacific that privilege egalitarianism and individual autonomy (although not necessarily diversity). For such societies, politics is carried out largely through talk, often various kinds of public oratory during collective gatherings. See in particular Atkinson 1984, Brenneis 1984, and Lederman 1984. I discuss the significance of talk as political discourse in this sense in chapter 5.

4. I distinguish here a discourse of national identity from one of nationalism that, conversely, seems to be strongest among working classes. Indeed, intellectuals are often known to revile nationalism with its emphasis on exclusion and its tendency to devolve into violence. In this way, discourses of national identity and nationalism are distinct, if related, phenomena and have different manifestations and effects. Scholars have grappled with this problem in di-

verse ways. For instance, Kapferer (1988) attempts to link these two concepts in his discussion of the ontological significance, the cultural force and meaning, of ideologies of nationalism. Anderson (1991) analyzes the development of "national identity" among the bureaucratic functionaries and administrators of colonialism, whose work led them across borders and helped to reinforce a sense of identity according to those boundaries. Bourdieu's (1984) discussion of "taste," moreover, explicitly links a valorization of and identification with "French culture" to class status and education. The political forces of European integration are now making the political borders of nation-states appear even in the most intimate settings (although as Sahlins 1989 suggests, this kind of penetration characterized state building centuries ago), suggesting that national identity and ideologies of nationalism are coproductions.

5. These efforts reflect an idea of culture as an object, something located in concrete symbols, rather than a more anthropological idea of culture as patterns and practices of meaning. The Europeans involved in this effort to catalyze a European identity might take stock of the attempts made by the Soviet government, in manufacturing symbols and even rituals, to offer its people a means of culturally and psychologically identifying with a state established primarily on political/ideological and economic grounds (see Lane 1981).

Nonetheless, as Varenne (1993: 236) notes, a meaningful Europe is being formed on the ground, as people must take into account such things as the ubiquitousness of the European flag (at gas stations, during press conferences by national politicians, and the like) and the proliferating use of ideographic signs on highways (which obviate the need for multiple languages). These things "are not mere curiosities. They are . . . concrete statements that historical persons have to [deal with] . . . in the current context." This, too, is Europe, a practical one, suggesting that "something major is being built that is transforming the stage on which private lives inscribe themselves" (ibid.: 237). In Zabusky 1993b, I have considered the question of the construction of a "European" identity among ESA professional scientists and engineers—some of these "experts" for whom Europe is supposed to be a meaningful entity—in more detail.

6. Historically in the West, science has been characterized by a vision of nature as female and the investigator as a male intent on "mastering," "dominating," and even "penetrating" nature (e.g., Keller 1985: 33ff.; Keller also notes the complexity of this imagery, since in the Baconian vision, investigators are also supposed to open themselves up "passively" to receive the wisdom of nature). There has also been a corresponding tendency in the West to equate men with rationality and women with irrationality culturally speaking (see for example Elshtain 1981). In an interesting complication of these cultural visions, European scientists conceive of nature as the domain of rationality, and they contrast it with the irrational vagaries of bureaucracy and politics. This view, too, has a historical and cultural pedigree, since Christianity's God, as Collins (1975: 294) reminds us, "is closest to the ideal of a rational rule-giver." It is striking to see that in this context of technology development (as opposed to scientific research, insofar as these can be distinguished), nature assumes a more masculine gender. One reason for the shift in this case, it seems to me, is

that nature appears as an ally for the scientists (and for the engineers, as well), particularly when they are arguing with "bureaucrats" or "managers." As such, nature is not something to be "penetrated" or "probed," but instead is an equal player and a bulwark against the (more female) ambiguities of social and political life. A deeper investigation of this significant opposition is beyond the scope of this book; I want to suggest simply that the genderedness of nature, science, scientists, and society shifts according to the larger context in which a particular set of practices is being deployed.

7. Latour (1987: 174) defines "technoscience" as "all the elements tied to the scientific contents no matter how dirty, unexpected or foreign they seem." In other words, he defines this term against normative, philosophical-oriented images of science as a hermetic activity dependent only on the work of scientists (excluding, for example, engineers, technicians, managers, bureaucrats, politicians, and the like) and only on the manipulation and negotiation of ideas (rather than of machines, equipment, money, social connections, and the like). I do not share Latour's overarching perspective, nor do I intend to take on his sense of irony or arrogance. Nonetheless, I adopt his term to convey a view that recognizes the interpenetration of science with technology, with macrostructures (such as nation-states and bureaucracies), and with other social actors (such as engineers and managers). The production of space science missions is not "just" science, in this sense, even if the satellites will be used by scientists to do scientific research at some point; "science" is an element in a more encompassing social and technological system of production (see Hughes 1987, Pinch and Bijker 1987, and Lambright 1994 for different but related discussions of the notion of "technological system"). Indeed, part of what I try to understand is how participants perceive "science"; in this respect, I am interested in the way participants articulate the boundaries around and between different elements of their practice (see also Cozzens and Gieryn 1990 for a discussion of different theoretical approaches to such boundary work).

8. Hagendijk (1990) decries also Latour's depiction of technoscience as part of a war machine; he argues that Latour tends "to reduce cognitive issues to the Machiavellian tactics of scientists acting as if they were political entrepreneurs or field marshals in combat. For Latour, science is warfare or politics continued with other means" (ibid.: 56). Amsterdamska (1990) also criticizes Latour's depiction of scientific "networks" for its Machiavellian cast. I share with these critics, and others, a distaste for the tenor of such analysis (see also Fujimura 1992).

9. Weber (1947: 137) might be a useful guide here. He reminds us that "no matter how calculating and hard-headed the ruling considerations in . . . a social relationship . . . may be, it is quite possible for it to involve emotional values which transcend its utilitarian significance. . . . Conversely, a social relationship which is normally considered primarily communal may involve action on the part of some or even all of the participants, which is to an important degree oriented to considerations of expediency." In other words, "conflict and communal relations are relative concepts" (ibid.: 138), and we err in privileging one over the other, or in reducing one to the other. Weiner (1976 and 1992) analyzes exchange relationships—the prototypical social integrator of

traditional societies such as those found in the Trobriand Islands where Weiner worked—as having just this kind of dual character; exchange relationships, moreover, characterize relations among scientists as well (Traweek 1988, Latour and Woolgar 1979).

10. After all, the trait of "sociality" (Howell and Willis 1989), as I discuss above, is about seeking out interactions and connections, not necessarily about harmony and agreement.

11. By "learn[ing] the community's cosmology and practices while doing fieldwork" (Traweek 1992: 439), I followed the typical pattern of anthropologists who choose to immerse themselves in such "on-the-job training" rather than enter degree programs or attend university lectures to become familiar with such material. Moreover, despite the limitations of my formal technical expertise, my own personal history endowed me with a deep familiarity with science, and this familiarity sometimes played a significant role in my relationships with participants. As the daughter of an engineer-turned-physicist, I had grown up around engineers, scientists, laboratories, and computers, and was accustomed to talking about things scientific and technological. As the wife of a physicist, I continued such association and conversation in my domestic life (neither my father nor my husband is, however, active in the fields of space science). These intimate details meant that I knew a lot about science in a general sense; it signified also, in a way, that even before coming to ESTEC, I was already affiliated with "the scientific community," or at least, as far as my informants were concerned, had a stake in it. The fact that I had such intimate connections to scientists and an interest in things scientific at times helped to allay participants' fears; indeed, I observed that sometimes when I informed them of these details, they were decidedly pleased, recognizing in me perhaps a kindred spirit, or at least someone whose eyes would not glaze over when they began to talk about scientific and technical matters. Importantly, this knowledge and connection validated me in their eyes much more than did my attendance at formal courses on astronomy. Thus, my own biography sometimes served as a methodological tool, facilitating conversation with participants in the field.

12. Another testimony to the interest people had in my study was the excellent attendance at a seminar I gave about eight weeks before I left SSD, at which I presented my preliminary findings. Although the biweekly SSD scientific seminars generally attracted audiences of about twenty-five people (and generally only scientists), mine attracted an audience of seventy-five, including engineers, technicians, and administrative staff (people with whom I had spoken during the preceding months) from inside SSD and from elsewhere in ESTEC.

Chapter Two
The European Space Agency and the Structure of Cooperation

1. The new convention stipulated that ESA would undertake the following: a shift of orientation from science projects toward applications satellites, in particular a maritime navigation satellite project; a launcher development pro-

gram, picking up where ELDO had been suspended, but replacing the old version (Europa) with a new one (Ariane); and a scientific project called Spacelab, which was to be Europe's contribution to the American post-Apollo program (Longdon and Guyenne 1984: 29). It is significant that science went from being the cornerstone and major activity of joint European space efforts to being one of many activities, and a small one at that (see Russo 1993a and 1993b). I discuss the role of the science program in the last section of this chapter. I have discussed the effects of this problematic status of science on the equally problematic professional identity for ESA staff scientists in Zabusky 1992. For details on the history of ESA, see Longdon and Guyenne 1984, European Space Agency 1989, and Battrick 1984, and the series of ESA History Study Reports (to date, ESA HSR-1 through HSR-8 and HSR-Special have been published; see in particular Krige 1992 and 1993 and Russo 1993a and 1993b).

I should also mention that not all space projects in Europe are undertaken under the auspices of ESA. ESA's work is complementary to, and does not supersede that of, individual member states, some of which have extremely active national space programs (Russo 1993b). ESA's omission of military-related space work itself necessitates the continuation of independent national programs in space technology, since this is not an area in which European nations have yet managed to come together in concerted fashion, although individual projects are being undertaken in a regional context.

2. Although the quotations I use in this section are drawn from material written in the 1980s, I use the present tense in my analysis, since I believe the cultural dimensions that I elucidate here continue to motivate and emerge from technical and organizational work in ESA.

3. In the text, I give examples of this kind of perspective in the discourse surrounding ESA. For an example of this "functionalist" perspective in the EC, see, for instance, the 1990 Intergovernmental Conference in the Context of Parliament's Strategy for European Union, as discussed by Holmes (forthcoming).

4. This is, at least, how the Europeans see it. It is not entirely obvious that the United States views such cooperation with its European counterparts in the same way. For instance, the Hubble Space Telescope mission, launched in April 1990, offered an example of cooperation between Europe and the United States in which the Europeans provided 15 percent of the budget, in the form of the solar arrays (an indispensable part of the satellite) and one of the four scientific experiments that the telescope carried (a faint-object camera, which will be available for use by all participating scientists, not simply the European builders of the camera). ESA understood in this venture that its participation was virtually nonessential, and that the United States could have afforded to carry out this project independently. It was the quintessential "little brother," allowed to tag along but not to play. That the United States perceived the Hubble Space Telescope as an *American* project was clear from the news coverage prior to and after launch, in which the participation of ESA was seldom if ever mentioned; even when it was mentioned, ESA appeared as a supplier of a subsystem, rather than as a partner.

In the case of the U.S. Space Station, however, ESA has conceived of its role as indispensable. Here again, ESA's portion is small, but (at least in the initial design of the project) it includes three integral components, in addition to contributions to the budget and the crew component, without which the United States could not proceed with project development. Other international partners include Japan and Canada, and, since April 1993, Russia as well. From the European perspective, this is an international project with the United States in the role of coordinator (i.e., a passive leader), a project to which each partner contributes what it can, up to its level of interest, and in which each partner is an equal. From the American perspective, at least as far as can be assessed from public pronouncements and policy actions, the space station is a project actively led by the Americans with some international assistance to help defray costs. This stance has led the United States to take unilateral action in design and scheduling changes in recent years, as the cost and magnitude of the undertaking have loomed larger and larger, much to the anger and dismay of its international partners, in particular, Europe. Again, in the U.S. media, discussion of the space station has rarely included any mention of its international partners, although this changed somewhat when Clinton took office in 1993, since his administration has placed a renewed focus on international cooperation as a practical means of insuring technical efficiency and cost effectiveness.

5. Austria and Norway entered into full membership only in 1987, although they had been associate members for some time before that. It is expected that Finland will also become a full member in the future. In addition to these nations, Canada has an arrangement with ESA for "close cooperation," permitting it to participate in certain optional (i.e., not scientific) programs.

6. The ESA budget is calculated in the AU (accounting unit), which is equivalent to the European currency unit (ECU), the currency used by the EC in its operations. The AU is a variable currency, and its value is calculated every year, based on the average exchange rates of the different national currencies over the month of June of the previous year. In 1994, 1 AU equaled 1.18495 U.S. dollars.

The specific amounts mentioned in the text have, needless to say, changed, although the relative contributions/allocations remain roughly the same. For instance, figures for the 1993 budget showed the total ESA budget at 3,156 million AU. Space Transportation Systems still received the lion's share of that budget, at 41.8 percent, and the mandatory programs had decreased to 18.6 percent. France still contributed the most, at 29.22 percent, and Ireland still the least, at 0.19 percent.

7. At the time of my fieldwork, the director general was Reimar Lüst, a German plasma physicist. He was succeeded in October 1990 by Jean-Marie Luton, the director of CNES, the French national center for space studies, and a geophysicist. Luton's appointment has been extended twice, through October 1998.

8. In 1983, a total of 1,357 persons were employed by ESA. Beginning in 1985, a new recruitment program got under way in response to the initiation of a new long-term program, a program that included ESA's commitment to the U.S.-led Space Station *Freedom* project. By 1993, the staff complement had risen even further, to 2,064.

9. In addition to these ESA sites and various satellite tracking stations belonging both to ESA (for instance, at Redu, in Belgium) and to other national space agencies, there are numerous smaller establishments that carry out ESA-related work. For instance, there is the newly created (in about 1991) European Astronauts Centre for training astronauts to participate on Space Station missions (located in Cologne, Germany), and a department dedicated to preparing satellites for launch on Ariane rockets (based in Kourou, French Guiana).

10. According to a 1983 document (European Space Agency 1983: 11), "in real terms, ESA's scientific budget has remained virtually constant for the past decade." Currently, "science demands no more than ten per cent of the budget" (Russo 1993b: vi; for details on these historical developments, see also Russo 1993a).

11. Science is also significant in the Agency for larger political reasons. It is important for ESA to participate in space science research because this is widely perceived as a component critical to acquiring high prestige in the international arena. For Europe as a region to remain competitive with the United States, and, until recently, the Soviet Union (the heretofore acknowledged leaders in all domains of the space field)—and now Japan as well—it must maintain a visible level of involvement with space science in order to support its claim to being a space superpower. That is, for Europe to succeed in its bid for superpower status, it is not enough simply to participate in a space agency that brings pragmatic rewards; the region must demonstrate its ability to compete in this (apparently) altruistic endeavor as well, supporting science research that is perceived as being good for humankind, and not just for the particular economic interests of sponsoring governments. The commitment and ability to fund programs of scientific excellence thus confer on ESA a certain legitimacy as a leader in global politics.

12. This increase, because it came in the mandatory budget, required the unanimous approval of all member states. Although all agreed to the substance of Horizon 2000, Britain blocked the increase in the budget for two years, unwilling to augment its already sizable contribution. This stalled efforts to reanimate the Science Programme. After heated debates and angry words, including the widely circulated rumor that Britain would be expelled from the Agency, the stalemate was finally broken in 1989 with a compromise agreement. (See Berry 1988 and Fagan 1988 for newspaper reports on Britain's recalcitrance.)

13. The original Horizon 2000 plan also called for studies of "green dreams," missions that were beyond contemporary financial and technical capabilities of the Agency. These have been overtaken now by a "Post–Horizon 2000 Survey Committee" to investigate the parameters of future missions. Note, too, that the proposed launch dates for specific missions that I provided in the text are accurate as of spring 1994. These dates often change, however, as unforeseen events overtake the mission schedule. I include them primarily to give some sense of the time-scales involved in planning space science missions.

14. Sometimes it can take much more than five years to get a mission launched, depending on the vicissitudes of the space business. For example,

the explosion of the space shuttle *Challenger* in 1986 resulted in the delay of numerous missions, both American and European. One of these, Ulysses, already delayed three years, was further delayed another four and a half years. By the time it was launched, in fall 1990, it had been almost twelve years from the inception of Phase B to the beginning of operations. These delays represent more than just time; they represent as well significant cost increases in the development of the missions.

Chapter Three
The Practice of Cooperation: Working Together on Space Science Missions

1. The opposition I discuss here between the technical workers and the administrative and political workers on ESA missions reflects an opposition common to most bureaucratic organizations: labor and management are opposed, each seeing itself as responsible for the success of the institution and the other as responsible for putting obstacles in the way of that success. What is distinct in ESA is the particular way in which this opposition was perceived, in terms of the rational and pragmatic activities and orientation of technoscience versus the irrational and cynical activities and orientation of politics and the bureaucracies it establishes. This version of the opposition was itself contested, however; see chapter 5 for a more complete discussion of the politics of categorization in working together.

2. Also on the grounds, since about 1991, is the Noordwijk Space Expo, a museum about and showplace for European space missions and technology. Although on ESTEC grounds, just inside the main gate, it is run by an independent foundation.

3. This was the case during the time of my fieldwork, 1988–1989. The details I describe in this section reflect the situation at ESTEC and in SSD at this time. Since then, some things have changed. Just as I was leaving ESTEC, engineering and administrative staff in the main building began to move into the offices in the new tower block. At the same time, a relocation (within the original building) and renovation was planned for the offices and laboratories of the Space Science Department. Thus, the layout I describe here is no longer to be found.

4. The office towers stood between the original ESTEC office building and the new building housing the library, canteen, and conference center. These three structures were connected by a ground floor corridor. This corridor housed ESTEC's official reception area, the Space Store (which sold ESA souvenirs of various kinds), and a wide lobby in which professional artwork was regularly displayed. During most of the year that I was in residence, however, the new offices including the connecting corridor were still under construction; as a result, it was necessary for people to go outside and walk along the road that wound through the ESTEC grounds to enter the canteen.

5. Two exceptions to this were the Publications Division and the mock-up of the Columbus Attached Laboratory. The Publications Division was responsible for publishing a wide range of material, including all of ESA's technical

monographs, annual reports, proceedings of ESA-sponsored conferences, the ESA *Bulletin* (a quarterly publication with general information on ESA activities), and the ESA *Journal* (a quarterly publication containing papers on technical topics). The mock-up belonged not to the Technical Directorate but to the Space Station Directorate; it was the department in which technical staff developed and tested new technology to aid the astronauts who would work in the Laboratory during in-orbit flight periods.

6. As an indication of how hard it was to find office space for new incoming staff, take my own situation. I was provided with a desk, a telephone, and a computer terminal; this setup was located not in a staff office, however, but in a Department conference room. For the first four months of my stay, I was periodically asked to vacate my "office" so that other staff members could hold meetings in this room, a long, narrow room whose windows looked across a rooftop into other offices in a parallel wing. In January, following the winter holiday break, two research fellows were moved into my "office" as well. From this point on, the room ceased to be treated as a conference room and became one more SSD office, shared by three temporary people. The new office towers, although intended to relieve this kind of crowding, proved to be insufficient to meet the demand for office and meeting space, as the number of staff at ESTEC continued at that time to grow.

7. Traweek (1988) noted that such public bulletin boards were also present in the halls of SLAC, the Stanford Linear Accelerator; she never saw anyone but herself stop to read the material on them. Although, as I indicated here, this was generally true in my case as well, the bulletin board outside the Astrophysics main office proved to be an exception. On this board, the division head often made a point of posting some particularly absurd or outrageous article or comment pertaining to space-related issues, highlighting the interesting or offending passages in magic marker. This was not, strictly speaking, information; it was a kind of self-presentation, an announcement of a disciplinary and occupational identity of distinction within an otherwise homogeneous crowd. Others beside myself could often be seen perusing this commentary.

8. Sometimes people would speak of such differences in larger (interstate) regional terms, as when they spoke of those from "the North" (meaning Scandinavia, Germany, the Netherlands) who were unlike those from "the South" (Italy, Spain, and sometimes France). In this case, "the North" was the locale of dourness, formality, and order; "the South" was the locale of passion, informality, and unpredictability.

9. These staff engineers are based in the specialized divisions of the Technical Directorate (for instance, the attitude control division, the quality assurance division, the thermal division, and so on). Most of the engineers working for the Technical Directorate "generally work in a task-sharing mode" (European Space Agency 1985: 48), in which an individual may be asked to work on a variety of different missions (not only scientific ones) at any given time. This organization of work reflects the modified matrix structure that characterizes the bureaucratic structure for ESA's technical work.

10. One of the staff scientists said to me, after I had observed a working session between him and a couple of SSD technicians, "You could write several

dissertations about the relationship between technicians and scientists in SSD"; the same is true for that between technicians and engineers. This book is not, however, about SSD, as I explained in chapter 1. The tense relationship between technicians and higher-grade staff is a significant factor in the work of SSD as a department of ESA, but it is less significant in contexts of working together on space science missions. I do not mean to suggest that this conflict is unimportant, simply that it would have to be the subject of another book (or two).

11. The Earth Observation and Microgravity Directorate also, to the best of my knowledge, hires scientists to work on ESA projects in these two domains. The Microgravity program was, however, still relatively new and quite small at the time of my study, and its staff members were located in the Paris Headquarters. From the perspective of the staff scientists in SSD, the staff members of the Microgravity program were not scientists of the same order, because their interests did not pertain to space itself but merely to using the environment of space to conduct biological experiments like those that could be conducted on earth. The two groups of scientists thus had little in common and absolutely nothing to do with each other at this time.

12. This scenario represents a change implemented during the time of my field stay in SSD. Formerly, staff scientists were awarded a series of four-year contracts, never knowing from contract to contract whether they would be asked to continue. In this way, some staff have ended up working in SSD for as long as twenty years, but always with the uncertainty of these short-term contracts. In 1984, certain administrative changes were made in the Agency overall that affected the engineering staff, giving them indefinite contracts after an initial four-year probationary period, but did not affect the "rotation policy" applied to the scientific staff in SSD.

13. The use of these terms shows also the particular way in which participants codify and reify their activities. "Scientists" and "engineers" are, in other words, less different people than they are different modalities of a common process. All participants may, at one time or another, engage in the kinds of activities or exhibit the kinds of attitudes associated with "science" or "engineering." The flexibility of practice in this regard, however, is not relevant in the particular social calculus of working together, in which social categories of difference identify individual *participants* (and groups of participants) rather than widely available cognitive activities. Yet even these identifiable social actors slip out of such categories; in chapter 5, I show how people manipulate these terms and their definitions. Here, I simply lay out the ideal types as they are articulated in everyday discourse.

14. Not all Science Teams have such mission scientists, although it is becoming an increasingly common practice for ESA to include these. They are selected through a competitive process, as are the PIs and the instruments they represent.

15. Because of my affiliation with SSD, I had little or no contact with industry engineers, except on those occasions when they came to ESTEC to attend design reviews or other meetings.

16. As I argued in chapter 1, this process toward increasing and encompass-

ing integration resonates in intriguing ways with the logic of European integration: the latter also structures itself as a hierarchical order in which higher levels integrate constituents at lower levels, which themselves are composed of diverse smaller components.

17. In chapter 7, I return to the idea and activity of technological integration, and I discuss the metaphorical and symbolic implications of the technological outcomes that result from the working-together process.

18. Although linguistic abilities varied from person to person, there was a belief among participants that such ability varied from national group to national group as well. For instance, the French were notorious for having terrible accents, being unable to pronounce things correctly even if they were reasonably fluent. During one design review I attended, at which many engineers from a French aerospace company were giving presentations, an SSD scientist leaned over to me to whisper that "the French always have such terrible accents; you cannot understand them." This was considered part of the French "mentality," which viewed its own language with (to some, intolerable) pride. Italians were also considered to have difficulty in learning to speak English well, and to prefer speaking their own language. The Scandinavians, on the other hand, were widely viewed as being able to speak English as well as their own mother tongues; I found this to be true in my own experience.

Chapter Four
Struggling with Diversity: The Social and Cultural Dynamics of Working Together

1. Both of these statements about making social groups are also statements about political control or the lack thereof. The value of diversity corresponds to a positive valuation of equality; in working together, people resist the establishment of social hierarchies of control. I address this issue in detail in chapter 5.

2. The names I give to people in this excerpt and subsequent ones are pseudonyms. In order to highlight the way in which participation in talk is significant, I present here an edited excerpt of the longer argument (which lasted about twenty minutes, with a lengthy digression in the middle to discuss some technical matters relevant to the project in question).

3. This reference to "salmonella," which conveyed the sense of distaste (and danger) that Nigel was expressing, alluded to an earlier aside in the conversation, when people were discussing the recent (1989) scandal in Britain regarding the widespread contamination of eggs with salmonella.

4. The following discussion is a paraphrase from my field notes of what this person actually said over lunch.

5. This is typically where I sat during meetings. In Science Team meetings, others sitting at the margins might include engineers from the Project Team or from industry, or other SSD scientists not officially part of the Science Team. This was not a general rule, however; sitting at the margins represented nonparticipation more than lack of official inclusion, and even this was not always strictly the case. I attended meetings where, for instance, I was encouraged to

join the group at the table, and where SSD scientists not formally part of the Science Team sat at the central table and participated actively in parts of the proceedings. I, however, never participated, except to laugh at jokes.

6. This reflects, again, participants' Durkheimian view of their own process. Meetings are supposed to be like those rituals of aggregation when "thoughts are centred upon their common beliefs, their common traditions, the memory of their great ancestors, the collective ideal of which they are the incarnation; in a word, upon social things" (Durkheim 1915: 390). In this case, those "ancestors" are prior technical decisions, and the "social things" are embedded in the relationships that different groups (and their representatives) have to distinct technological components.

7. Laura Nader (1990: 291), in her recent discussion of "harmony ideology," encourages analysts to break harmony down "into its various components in order to understand its meaning and controlling power." Citing work done by Rose, she discusses the "multidimensional nature" of harmony in the Swazi case: "unity, consensus, cooperation, compliance, passivity, and docility" are all included in the meaning of harmony (ibid.: 296). I am here exploring just this kind of multidimensionality, which is, in fact, a source of the ambiguity that participants experience in relation to this value. I am not, however, arguing that harmony is a "controlling" ideology, which is one aspect of Nader's argument. "Harmony" in the case presented here is instead a value, a cultural trajectory inside practice. What I call "cooperation" more nearly resembles the "harmony ideology" of which Nader speaks. I discuss the productive links between harmony and cooperation in chapter 7.

8. The contrast I draw here between harmony talk and conflict talk resonates with a distinction made by F. G. Bailey (1965: 10) between different types of decision-making councils: elite councils and arena councils. In elite councils, participants attempt to "damp down dispute" because "they have a strong incentive to present a front of consensus and keep their ranks closed." Members of arena councils, on the other hand, are "steered by the heavy rudder of those whose interests" they represent. I argue here that Science Teams and other such social groups in the working-together process cannot be straightforwardly defined by what type of council they are, especially because such social groups are themselves always coming into being, even in the moment of their manifestation in meetings. They take on different attributes at different moments; sometimes they seem to be elite councils, other times arena councils. It is important therefore to pay attention to the contexts of specific interactions in order to understand when the "dominant cleavage" in a (temporary) social group is "horizontal" (elite) and when "vertical" (arena). This is why I refer to different kinds of talk, rather than different kinds of groups.

Chapter Five
Evasion and Responsibility: The Politics of Working Together

1. Moore (1978: 55) defines such a social field as one that "can generate rules and customs and symbols internally, . . . [it] is also vulnerable to rules and decisions and other forces emanating from the larger world by which it is surrounded." In the context of working together, however, such social fields are

not, as I indicate, "surrounded" by others, a term that conveys an image of ever-widening concentric circles such that interior social fields are embedded in exterior ones. In working together, there is no clear-cut interior or exterior; the social fields cut across one another, creating areas of overlap at the different points at which they intersect.

2. This kind of attitude will not be unfamiliar to anyone who works in academic circles. In universities, for instance, faculty tend to be as hostile to the "administration" as are these SSD scientists, perceiving the management issues that preoccupy senior administrators as so much "dirty work," and budgetary and policy decisions as potentially interfering with their scholarly and pedagogical tasks. Similarly, hostility to "management" on the part of labor is endemic to organizations in industrial capitalism.

There is another dimension to this hostility in this particular ethnographic case, however. Headquarters stands not only for "administration" but for "France"; many regarded the French as having an undue interest in running ESA, and indeed as having undue influence as well. Stories circulated about French government officials forcing French scientists to lobby for scientific missions that would enhance their country's status in space science, and allow the French government to play out its own political agenda, independent of considerations of scientific and technical merit. Significantly, non-French scientists and engineers did not project this hostility against "the French" onto their colleagues with French citizenship; French scientists and engineers actively involved in the technical work of space science missions were viewed as distinct from French administrators and politicians. As I have discussed above, "nationality" is relevant only in certain contexts and is typically de-emphasized in contexts focused on work rather than politics, a significant distinction in the view of participants.

3. In this example, my use of the term *cooperation* is not analytic but decidedly emic; it should be read as part of the interviewee's discourse. For this administrator, the term *cooperation* referred primarily to the Agency's ability to bring together the interests of diverse nations on space matters. This was political-economic cooperation, not practical-technical cooperation; her use of this term also did not address the way in which people worked together on a daily basis.

4. I think also that her reference to "farmers" had another connotation, emanating as it does from the "cultural" center of Europe—Paris—and looking toward the bucolic periphery of Holland. Many Europeans consider Holland to be a country of farmers (in Dutch, *boeren*, pronounced "boor-e"), a country without class or sophistication, home to the petty bourgeois and their country cousins (see Schama 1987 for a historical perspective on this view of the Dutch). In a way, this administrator was thus referring to the "boorishness" of scientists and engineers as well, conflating the agricultural Holland with the political naïveté of technical staff.

5. Dumont (1986: 252) discusses the way in which symmetric oppositions of values are rendered meaningful only in a hierarchical relationship, in which one value "encompasses" the other, such that "'high' ideas will both contradict and include 'low' ideas."

6. This observation was not without foundation. A visiting American engi-

neer who had long worked for NASA also described to me how NASA was run in a much more authoritarian way, in accordance with a social hierarchy embedded in bureaucratic principles, than was ESA.

7. Although the scientists I knew would deny the relevance of hierarchy to their own work (itself a kind of mystification, since it obscures the way in which the meritocratic principles of the scientific community organize people in a social hierarchy; see Traweek 1988 and Zabusky n.d.), they did not thereby romanticize the egalitarianism of their community—as does, for example, Hyde (1983) in his depiction of the scientific community as a "gift community." They often talked about, for instance, the intense competition for prestige that resulted from the absence of a definitive status hierarchy. This competition was largely conceived of as a competition among equals, but it was one that often bordered on the destructive.

8. It is interesting to note here a confluence between the practices of working together and the analytic concept of *cooperation* as it is defined by game theorists (e.g., Axelrod 1984). In game theoretical cooperation, the formulation of successful cooperation is precisely that everyone wins. What everyone wins is less than what one person might win when conquering an adversary, but such higher winnings are at the expense of the other person (who gets nothing), such that the system itself comes out with a lower score. Successful strategies are those which enable everyone to get something by consistently cooperating with each other, such that the system itself comes out ahead.

9. My depiction of the meaning of the terms *dictatorship* and *democracy* here reflects participants' cultural views. I am not making an argument about the political ideals and political realities of any actual or theoretical governmental systems based on these principles. Nonetheless, that participants think about their social order in these terms reveals that they are not focused simply on how to make the most efficient and most rational technical and scientific decisions. They share in the wider concerns of European citizens in a postwar world who are trying to figure out how to establish social polities that will permit freedom and equality for all members. The nation-states of Europe have experimented with a variety of forms of democracy, particularly the liberal democracy of the social welfare state. The federating EC is another such experiment in the extension of individual rights through the potential abdication of sovereign authority. It is just this sort of political experiment in which participants in working together are also engaged. (Both of the experiments I mention here, of welfare states and European community, have been criticized as being simply old wine in new bottles—that is, as new versions of the institutionalized domination of the working class by capitalist interests.)

10. The situation was different, I was told, in Project Teams. Project Teams were constituted entirely by ESA staff members; as such, their members' status and position were influenced by ESA's bureaucratic hierarchy. The project manager could, in fact, say "Jump" to the other engineers on his team. He had a more difficult time saying "Jump," however, to engineers in industry and to scientists and engineers working as members of instrument consortia, all of whom had independent sources of funding and budgets, as well as institutional hierarchies to defend their positions.

11. Again, I am not referring to meetings of teams or groups constituted entirely by ESA employees, such as SSD division meetings or Project Team meetings. In these contexts, a clear bureaucratic hierarchy and its accompanying location of authority is available. I refer instead to the diverse social groups that are formed out of the diverse participants in working together on space science missions.

12. Giddens (1984: 16) argues, in fact, that "all forms of dependence offer some resources whereby those who are subordinate can influence the activities of their superiors." This is the "dialectic of control." In this case, the "subordinate" position of the PIs was entirely relative, however, as I discuss in the text.

Chapter Six
Existential Worries: Excitement and Boredom in the Experience of Working Together

1. Such moments, as I argue more fully in the Epilogue, are essential in keeping the process of working together alive and moving, even if they are not—indeed, cannot be—sustainable.

2. ESTEC supports a wide range of social, athletic, and arts activities through the SSCC (the Sport Social Cultural Committee) and through ESCAPE, a recreation facility on the grounds of the physical plant, which opened in April 1988. The recreation building houses a small swimming pool, squash courts, saunas, a solarium, a fitness/weight room, indoor tennis courts, a sports hall suitable for basketball or football (i.e., soccer), a dance room, club rooms, libraries (of videotapes, music tapes, and books), and a bar that serves drinks and snacks. The building is open from early morning till late in the evening; staff members and their families are eligible to use it. I did not have regular access to these facilities, but participants' stories about football tournaments (where national teams were organized) and other such recreational activities made clear to me that it was in such "extracurricular" pursuits with colleagues that passions relating to nationality particularly came to the fore.

3. In a way, even these white-collar professionals can be considered migrants, even though their salaries put them in another class altogether from that group of people normally considered as migrants, and even though their movement to another country seems to take place in a context of freedom as opposed to the economic constraints that force the working-class migrant to work outside his/her native country, often leaving family behind for years on end. Although many ESTEC expatriate staff bring their nuclear families with them, this is not always the case; moreover, they are separated from other family relationships that in many instances are as significant for them as those of the nuclear family. Furthermore, at least for many of the scientists, taking up a position at ESTEC is not experienced as an act of freedom but rather as an act of desperation—not economic desperation, but personal desperation, because for many, ESTEC offers the last chance they have to continue to work *as scientists* before crossing over into another category or another domain (e.g., engineer in industry). For these individuals, multiple constraints force them into migrating to another country in search of the work that they want to do. (See

Buechler and Buechler 1987 for a discussion of working-class migrants in Europe.)

4. The difficulty of finding employment was sometimes due to legalities. Although any citizen of an EC country could take a job in another EC country without special permits, for those families who came from non-EC nations, such as Norway, Sweden, Austria, and Switzerland, special work permits were required. ESTEC took care of this for its own employees but did nothing to help spouses in this regard, since it was up to the particular employer to do so. Even for EC citizens, however, spousal employment was not straightforward because of differences in language and training. Few people spoke Dutch before arriving in the Netherlands, and different countries had different credentialing systems for the various professions, making employment outside the home country problematic.

5. As I indicated in the introduction to this chapter, I am not focusing on psychological states here, but on cultural conceptions of emotions. What I have been describing, then, is a particular cultural meaning of excitement, which is not necessarily universal. As Ries (1993) points out in her work on Soviet Russian society, "excitement" is an unfamiliar term and an unrecognized experience for most Muscovites. Similarly, Bateson (1972: 111–12), in his discussion of the "steady state" of Balinese society, indicates that the kind of excitement characteristic of schismogenic societies (including but not restricted to Western ones) is absent in Bali, where "noncumulative interactions" typify social relations. Bateson notes that the psychological state of excitement that we recognize in the West is part of the "struggle" engendered by schismogenic cycles, and it comes in particular from the "climax of involvement," which he likens to sexual orgasm. The relevance of this analysis to the one I am undertaking here is suggestive.

6. As I have argued above, scientists viewed the launch as the beginning of the work necessary to achieve mission goals, which focused on the collection and analysis of scientific data. For scientists, it was only when the satellite became operational that they could start collecting data. And it was only when the data were compiled that analysis could begin. From the perspective of ESA, however, data analysis was "post-operations"—an afterthought. At the time of my fieldwork, the Agency had allocated little money to this activity that was, for scientists, the most important phase in the schedule of a mission; the scientists were in the process of lobbying for increases.

7. Because the role of the schedule is common to all space science missions, it is a ubiquitous worry. Particular types of missions, however, are accompanied also by more specific kinds of time worries. On observatory missions, for instance, people worry about getting "observing time," for which most scientists have to compete. Those who work on the mission instrumentation, however, are given "guaranteed time," as opposed to the "open time" that is the focus of competition. Working on observatory missions means worrying about time all the time; during the sensitive negotiations over who gets how much of which kind of time to do what, tempers routinely flare.

8. Technology is designed, however, with a generous "margin," so that the designated mission duration time may be superseded. This occurred in the

case of the International Ultra-Violet Explorer, which survived long past its anticipated demise to mark its tenth anniversary of operation in 1989.

9. The 1984–1985 "Report on the Activities of the Space Science Department" (Page et al. 1986) reported that the average age of the scientific staff in SSD was 38.8 years, compared with an average age for ESA staff of 44.3 years; ages ranged from 30 to 51. The average stay for staff scientists in SSD at that time was 3.1 years (ibid.: 6). The staff scientists chafed against the policy of rotation, especially because it did not apply to any of the other ESA employees, particularly engineers, who received indefinite contracts after an initial four-year probationary period.

Interestingly, the policy of rotation reflects the presence of two value hierarchies ranked inversely according to age along which people move as their careers progress. There is a hierarchy of intellect, in which the apex is youth and down which people move as they get older. That is, the younger scientists are, the more creative and perhaps smarter they are, which gives them a certain moral authority on technical and scientific matters. Simultaneously, there is a hierarchy of control, in which the apex is maturity and up which people move as they get older. That is, the older scientists are, the more sophisticated they become and the more institutional power they acquire, which gives them authority over others particularly on bureaucratic and political matters (which are often conflated with technical ones as well). In other words, rotation keeps the people at the top of the hierarchy of intellect at the bottom of the hierarchy of control (which is also the organizational hierarchy), and vice versa, such that too much and/or too many kinds of authority do not come to rest in the hands of a particular cohort.

10. There are those who do not subscribe to the belief that they become less creative or intellectually capable as they age. One SSD scientist said to me that "experience" was in fact just as significant as youth to successful work in space science research.

11. In discussing these death stories, I am not arguing that the occupations of space scientist or engineer are characterized by actual physical risk to life and limb every day. There are certainly many occupations that put practitioners into more life-and-death situations (doctors, emergency rescue personnel, high-rise construction workers, and so on). The point here is that these narratives are ubiquitous, even if the risks they discuss are not. As an anthropologist, my interest is in what these kinds of narratives tell us about the social and cultural preoccupations of participants. An aspect of what I am arguing here is that working together is not, for participants, just about technical details, money, and bureaucracy; it is not "just" a political or economic process, driven by concerns for more and better funding or besting opponents in a political drive for power (even if it includes these elements; I discuss this in the Epilogue more fully). Working together also informs people's identities; it is a process replete with meaning and not just with parochial interest. In working together, people think about what really matters to them; it is an existential matter of life and death because in working together, people discover who they are.

12. Certeau (1984) also links life and work in his essay "The Unnamable."

He argues there that "along with the lazy man, and more than he, the dying man is the immoral man: the former, a subject that does not work; the latter, an object that no longer even makes itself available to be worked on by others; both are intolerable in a society in which the disappearance of subjects is everywhere compensated for and camouflaged by the multiplication of the tasks to be performed" (ibid.: 190–91). The "multiplication of tasks" is an analytic way to talk about people's experience of being "busy" all the time, which participants certainly were, as I have shown. Certeau goes on to say that "in our society, the absence of work is non-sense; it is necessary to eliminate it in order for the discourse that tirelessly articulates tasks and constructs the Occidental story of 'There's always something to do' to continue"(ibid.: 191). In other words, death represents the impossibility of continued work, work that in "our society," to use Certeau's open-ended term, defines the meaning of our lives.

Chapter Seven
Working Together Transformed: The Production of Technology and Cooperation

1. Disagreeing with my interpretation, one participant pointed out to me that this example also demonstrates, in his words, that "time is money." In his interpretation, it was the financial expenditures involved in fixing and dealing with unanticipated problems that motivated the hostility and the dramatic stances other participants assumed in this confrontation. That is, what the recalcitrant PI was loath to do was admit that he had managed things poorly, which would mean that he needed to commit more funds to correcting the problem that had cropped up. Instead, he attempted to get the project engineer to take the blame for the problem so that the project and/or other instrument consortia would have to foot the bill for the necessary corrective work.

Such economic concerns no doubt play a role, but they do not sum up what is happening. This kind of interpretation is reductionistic, simplifying all action, and with it all meaning, to economics. It reflects, thus, a view of social processes as being economically determined; it reflects, too, a view that human beings represent, fundamentally, "homo economicus," a view which denies that there might be cultural or existential meaning entangled in such pragmatic tactics. I do not reject such economic interpretations; however, I do insist that there is more going on than economic wrangling and concomitant "face-saving" in this and other examples of conflict in working together.

2. This enlivening of artifacts that I am describing is analogous to the Marxian concept of fetishization. Taussig (1980: 33) notes that although this concept has been applied particularly to the production of commodities in capitalist systems, it is a "probably universal tendency whereby any culture externalizes its social categories onto nature, and then turns to nature in order to validate its social norms as natural." Bourdieu (1977: 164), indeed, argues that such an externalization with accompanying disguise is a critical part of any and all social reproduction: "Every established order tends to produce (to very different degrees and with very different means) the naturalization of its own arbitrariness."

3. Actually, it is the penultimate integration. The final one is that of satellite with rocket (or space shuttle), but this is a temporary connection. After launch, the rocket falls away or is otherwise detached, leaving the (integrated) satellite to function on its own.

4. By "transcendence" here I do not mean to refer to a mystical sense of illumination, but rather, in the words of Turner (1977: 69), to "an intrinsic feature of the relations between the levels of hierarchically organized structures"; that is, I use it without "connotations of 'sacredness.'" Such sacred connotations do, however, color my use of the term in the Epilogue.

5. This made the presentation session both "important" and trivial, a "boondoggle," as were other similar convocations such as design reviews (see chapter 4). For details on the selection cycle, see the discussion of Horizon 2000 in chapter 2. I was not able to attend the working group meetings at which the missions were assessed and one recommended; neither were the SSD scientists responsible for coordinating the work of the relevant Science Teams able to attend these sessions. I did, however, attend the day-and-a-half-long presentation session, the substance of the ethnographic discussion here.

6. In the Epilogue, I pursue in more depth an interpretation of the cultural dangers and powers of these artifacts, dangers that are related to the contaminating effects of human beings in the production of these artifacts, and the power that comes from the association of these artifacts with an impersonal nature. Here I concentrate on issues of structure and practice in relation to these artifacts—specifically, the way in which practice makes such dominating structures provisional.

Epilogue
Sacred Cooperation and the Dreams of Modernity

1. Theology is defined (according to *The American Heritage Dictionary*) as "the rational inquiry into religious questions, especially those posed by Christianity." I am using the term *theology* here, accordingly, to suggest that my concern is with, generally speaking, the spiritual dimensions of social practice. In this sense, I am extending the use of this term in much the same way as social scientists extend the use of the term *hermeneutics* to refer to interpretation of texts in general, and not simply of scriptural texts. In other words, I do not mean to suggest that what I am concerned about is the nature of (a Christian) God. The Epilogue is, however, concerned with the shape of the sacred, as I explore how it appears and how it is perceived in this context. When I say that I am pursuing a theology of cooperation, then, I am saying in part that the Epilogue represents a more interpretive and less explicitly ethnographic or analytic exercise.

But the Christian connections implicit in the idea of "theology" are not, as it happens, beside the point; they are, indeed, very much *to* the point. The scientists and engineers working on ESA missions are heirs to a particular construction of the sacred, one that equates nature and reason, as I will show. This construction has a distinctly Christian pedigree; indeed, there are direct historical and cultural links between Protestantism and science, as many scholars have shown (e.g., Ben-David [1971], Merton [1968], Berman [1981], Tambiah

[1990]). Scholarly understandings of the particular contours of this conception have been influenced by Weber's (1958) interpretation of the ideas of the calling and of salvation in *The Protestant Ethic and the Spirit of Capitalism*. For this reason, it is no surprise to find Weber's depiction of the ascetic Protestant striving for the *"unio mystica,"* for the experience of "a real entrance of the divine into the soul of the believer" (Weber 1958: 112), shadowing the text I write here.

It is not my intention, however, to carry through an extended analysis of these links, nor am I trying to make a causal or explicitly historical argument here. I am more concerned with explicating the cultural logic at work in the contemporary practices that I observed. Nonetheless, I occasionally comment on the parallels between ESA participants' views and those of Weber's ascetic Protestants in the chapter notes, in part to show that there is a shared cultural logic at work in these specific ideas about the sacred, community, and purity. In making these connections, I turn not only to Weber but, from time to time, to other ethnographic interpretations of similar views of the coincidence of nature and reason, views shared by people in distinct historical and geographic contexts such as Southern Baptists in an Atlanta suburb (Greenhouse 1986); Australian nationalists (Kapferer 1988); American particle physicists (Traweek 1988); and American scientists debating human rights issues (Zabusky n.d.).

2. These oppositions suggest what the academic scientist quoted in the text meant when he distinguished a "human problem" from a "real one." He implied that the problems in getting research done were, in the first place, "trivial." Moreover, they were artificial, which meant that they could be changed. "Real" problems could not; such problems included the lack of naturally occurring materials with which to observe astronomical objects in the ultraviolet wavelength.

3. This free community of equals is one face of the cultural conception of the "scientific community." As Dumont (1970: 265) has argued, however, the ideal of equality, characteristic of so many Western ideologies, operates "within a group which is hierarchized in relation to others." Segal (1991: 9), drawing on Dumont's analysis of hierarchy, equality, and race, notes that "equality . . . operates within groups defined in relationship to subordinates who are figured as social and biological outsiders." Thus, the universal equality of the scientific community masks the exclusion of people who are conceived as "biologically" different (in this case, as I discussed in chapter 1, people of color and women). This ideological effect of the ideal of equality, however, does not render its aspirational dimension irrelevant, as I argue here.

4. Kapferer (1988) describes Australian egalitarian ideology in much the same terms. He discusses the concept of "mateship" as one that is "at once expressive of . . . [egalitarianism's] ethos and a central principle of social coherence." Mateship is understood to be "the basis of natural society, the way society forms, independent of artificial mediating institutions such as those implicit in the concept of the state" (ibid.: 158). That European scientists and Australian nationals should share such a similar view of the value of their own communities is significant, as I shall argue in the conclusion of the Epilogue.

5. This configuration of sacred cooperation—the kind of relationships it imagines and the kind of community these produce—resembles in many ways the communitas of which Turner (1967, 1974) writes. Communitas is an idea of, in Kapferer's (1988: 162) words, "the oneness of humanity," which expresses "the character of a fundamental indivisibility of humanity." This concept applies so neatly to these contexts of European space science precisely because, as Kapferer argues, it "is probably most applicable to modern, nontribal societies . . . [particularly] those founded in ideological variants of egalitarianism" (ibid.: 163–64), even though Turner first identified communitas as an aspect of African tribal societies.

6. As I suggested above, and as I have argued elsewhere about another group of scientists (Zabusky n.d.), the scientific community is often imagined as such a sacred community, in which free and equal relations prevail, and in which such relationships are indeed made possible by and legitimated in the search for nature's objective truths. The shape of this sacred community, as I suggested in n. 1 above, resonates with the community of the saved imagined by Weber's ascetic Protestants. I have heard these resonances, too, in Greenhouse's (1986) discussion of a more contemporary Protestant community, that of some Southern Baptists living in an Atlanta suburb. The cultural parallels are striking. Greenhouse shows that these Baptists define the community of the saved as "being harmonious a priori by virtue of having accepted Jesus" (ibid.: 105). This view of community has implications for an understanding of conflict: "Just as Hopewell Baptists experience their salvation in part by transcending conflict, they experience the secular world as conflict that they must transcend" (ibid.: 119). Conflict is a (necessary, inevitable) part of the secular world for these Baptists, but it introduces the possibility of "winners and losers or—more generally—authorities and constituents" (ibid.: 105), and thereby the possibility of a social hierarchy. Since "not only is human authority inappropriate, it is also illegitimate," such hierarchies must be rejected within the community of the saved.

European scientists, too, identify harmony (or unity) as part of their sacred community, placing conflict on the outside, as part of the human world, as an effect of the contaminating passions that prevent people from achieving equality. They, too, see human authority as illegitimate (as I showed in chapter 5), trusting only nature to dictate action. As the Baptists see themselves as equal before God, so the scientists see themselves as equal before nature (where what is equal trembles in a confusion with what is the same, again leading to the exclusions that communities make).

7. In their commitment to "real work," participants appear—as my use of the term *vocation* suggests—much like the *Berufsmenschen,* Weber's ascetic Christians for whom "everyday worldly activity [had] a religious significance" (Weber 1958: 80). This is the meaning of the calling, work that is undertaken as an end in itself, as a form of meditation on the sacred (for the Puritans, the sacred resides in God; for engineers and scientists, it lies in nature). Weber shows how the ascetic Protestant sects understood work as being the best way, the only way, people had to live "acceptably to God" (ibid.). This was not ordinary work, even though it was carried out in the world rather than behind

the walls of a monastery, an exclusion that these Christians rejected. The work of the calling was preoccupied with proof, with the application of rational rules to a disorderly world, in order to permit an individual to lead "an alert, intelligent life" (ibid.: 119); thus, the effort to work only for the glory of God gave "a peculiarly objective and impersonal character" (ibid.: 109) to the work in a calling. As I discuss in the text, the scientists and engineers who dedicate themselves to the rigors of "real work" also find themselves fighting against profane contaminants (such as those subjective and "irrational" factors located in national politics, bureaucratic policies, and personal idiosyncrasies) in their effort to maintain the focus on rationality in the execution of daily tasks.

8. Restivo (1988: 215–16) offers a cogent critique of this "myth of purity." He discusses its articulation particularly in the world of mathematics, in which the "ideology" of purity, which is analogous to what I have identified here as "real work," "grows in large part out of ideologies of God and Nature as ultimate authorities" (note here again the connection with the cultural specificity of a Christian worldview). He argues persuasively that "reified realms of purity such as logic can be functional equivalents of God and can serve as *moral* imperatives and constraints that in one way or another bind us to established professional and state interests and reinforce obedience at the expense of criticism and rebellion in our relationships within established institutions" (ibid.: 216; emphasis in original). There is no denying this ideological effect of a belief in purity; however, purity operates also as an aspiration, as I argue in the concluding section of this Epilogue.

9. Other scholars have identified the significance of a dichotomy between real work and other kinds of work in the world of scientists and technologists. For instance, Forsythe (1993) shows how researchers in artificial intelligence ("knowledge engineers") distinguished between "work" and "other stuff," and MacKenzie (1987: 198–99) discusses the frustration guidance engineers felt when they had to pay attention to social, economic, or political matters instead of technical ones. Certainly academic social scientists, too, operate with such a dichotomy, often denigrating teaching and administrative duties as "other stuff" that impedes their (our) ability to focus on the "real work" of research.

10. Sometimes the acronyms themselves were playful, or made words with which participants could then make jokes. This tendency, particularly strong among the scientists when it came to naming their missions, was resisted by ESA's leaders, who issued a decree during the time I was in residence that henceforth, abbreviations for mission names must *not* spell words. The fact that there was some play in the system is indicative of the complexity of this practice. People's orientation to work was now pragmatic, now sacred; both perspectives arose in the same social space and in reference to the same resources and materials.

11. For a discussion of the links between national and occupational identity in work and leisure contexts at ESTEC, see Zabusky 1993a. For discussions of the way in which technical work focuses on the sharing of expertise at the expense of institutional or professional hierarchies, see Barley 1993 and Scarselletta 1993.

12. This stripping down and cleaning up that I describe here as being part

of "real work," and hence as a kind of sacred duty of objectivity, is analogous to the "culture of no culture" characteristic of the "community" of American particle physicists, as analyzed and critiqued by Traweek (1988).

13. This dream of transcendence shares many elements with the idea of the *"unio mystica"* and the desire for salvation that motivated the ascetic Protestants Weber wrote about. Greenhouse (1986), writing about Southern Baptists in the United States, also talks about the idea of salvation. Greenhouse's depiction of this idea corresponds to the dream of sacred cooperation that I depict in the text. For the Southern Baptists, "the idea of salvation takes shape as a yearning, as desire. . . . Salvation is . . . desired as a form of freedom . . . [particularly] freedom from the constraining judgments of other people. . . . Freedom is not the same thing as the desire for freedom. . . . In framing a desire for freedom . . . , believers articulate relationships, the idea of whose rewards they cherish" (ibid.: 197–98). Participants in working together also cherish a particular vision of social relationships, as I have indicated; they, too, recognize that desire is not the same thing as the object of that desire.

14. Many nationalist ideologies, recognizing the need for "pure" communication in the pursuit of such quasi-religious commitments, try to purify (national) language as well. Herzfeld (1987: 52) discusses the efforts of Greek scholars to use a "Greek purist language"; this effort reflected "the romantic vision of a scientific state . . . matched by a scientific rational language. Language, no longer dependent on the exigencies of context, would now become an absolute and precisely calibrated tool," equivalent to the language of mathematics in science. Herzfeld's linking of science to the state, here, further underscores the linkages in this ideological arena that I am trying to bring out in this section.

15. Brenneis (1994), in an intriguing ethnographic study of the processes involved in American social science peer review, shows how this "purifying" effort of bureaucracy works in practice. He describes how participation on evaluation panels "plunges one into . . . a normalization process." This process depends on "self-discipline," as participants are encouraged to, or are seduced into, "acquir[ing] a new language" that emphasizes criteria of standardization, clarity, and comparability. These criteria bring reviewers to evaluate proposals through a lens that filters out idiosyncrasy—bureaucrats, even more than scientists or engineers, insist on stripping themselves down in the production of "fairness."

16. The specific contours of this idea of "rationality" once again resonate with beliefs held by Weber's ascetic Protestants, who made of rationality a religious dogma. In its Protestant form, rationality was primarily an instrumental tool. It represented "the technical control of . . . chances in the world . . . without regard for its effect on the (substantive) rationality of the larger world" (Eisen 1987: 112). Rationality was, for these Protestants, a tool of purification, insuring salvation as it protected believers from the contaminations emanating from "the degradation of the flesh" (Weber 1958: 153).

17. I am not denying that rationality operates as an ideology in the cultural space of modernity. What I am arguing is that the purity which rationality promises is not *only* an ideology. When people *claim* to have achieved this

purity, it is ideology; when people *regret* that they cannot achieve it, it is, instead, aspiration. This is a critical distinction.

Claims of purity, of exercising rational and unambiguous action unaffected by the "noise" of context and subjectivity, are legitimizing claims articulated in the struggle for control. They can lead, accordingly, to the domination that participants fear, when purity is reified in structures that deny subjectivity and difference. This ideological purity also leads, importantly, to oppressive and discriminating exclusions, exclusions of which participants seem to be unaware, caught up in the collective forgetting that is the hallmark of ideology (Bourdieu 1977). But participants also have regrets, as I have argued here. Regrets are the flip side of desire and signal the presence of aspirations—in this case, the striving for moral links among human beings engaged with each other and with the world. This is an aspiration for purity, in which purity is best understood as a moral ideal, such that it imagines "a better or higher mode of life . . . where 'better' and 'higher' . . . offer a standard of what we ought to desire" (Taylor 1992: 16). I call it an aspiration in order to call attention to the unattainability of this desire, and to the collective recognition that it must be striven after, that it cannot be taken for granted.

Claims and regrets are inextricably linked, although they are not the same thing. I suggest that in recognizing the legitimating dimension of ideological claims, we should not blind ourselves to the liberating possibilities of whatever aspirations are also implicated. These are, inescapably, the source of our own basis for critique; in other words, there is room, indeed, there has to be room, for improvisation, creativity, and play even in the spaces of power.

References

Amsterdamska, Olga. 1990. "Surely You Are Joking, Monsieur Latour!" *Science, Technology, and Human Values* 15 (4): 495–504.

Anderson, Benedict. 1991. *Imagined Communities: Reflections on the Origin and Spread of Nationalism.* Rev. ed. London: Verso.

Atkinson, Jane Monnig. 1984. "'Wrapped Words': Poetry and Politics among the Wana of Central Sulawesi, Indonesia." In *Dangerous Words: Language and Politics in the Pacific*, edited by D. L. Brenneis and F. Myers. New York: New York University Press.

Axelrod, Robert. 1984. *The Evolution of Cooperation.* New York: Basic Books.

Bailey, F. G. 1965. "Decisions by Consensus in Councils and Committees, with Special Reference to Village and Local Government in India." In *Political Systems and the Distribution of Power*, edited by M. Banton. A.S.A. Monograph. London: Tavistock Publications.

Barley, Stephen R. 1991. "Contextualizing Conflict: Notes on the Anthropology of Disputes and Negotiations." *Research on Negotiation in Organizations* 3:165–99.

———. 1993. "What Do Technicians Do?" EQW Working Paper. Philadelphia: National Center on the Educational Quality of the Workforce, University of Pennsylvania.

Bateson, Gregory. 1972. "Bali: The Value System of a Steady State." In *Steps to an Ecology of Mind.* New York: Ballantine Books.

Battrick, Bruce. 1984. "ESRO + ELDO = ESA: The Men and the Milestones." *ESA Bulletin*, no. 38: 20–30.

Ben-David, Joseph. 1971. *The Scientist's Role in Society: A Comparative Study.* Chicago: University of Chicago Press.

Berman, Morris. 1981. *The Reenchantment of the World.* Ithaca, NY: Cornell University Press.

Berry, Adrian. 1988. "Euro Space Agency May Oust Britain for Not Increasing Fee." *Daily Telegraph*, 17 October, 22.

Bloor, David. 1991. *Knowledge and Social Imagery.* 2d ed. Chicago: University of Chicago Press.

Boerner, Peter, ed. 1986. *Concepts of National Identity: An Interdisciplinary Dialogue.* Baden-Baden: Nomos Verlagsgesellschaft.

Bohannon, Paul, ed. 1967. *Law and Warfare: Studies in the Anthropology of Conflict.* American Museum Sourcebooks in Anthropology. Garden City, NY: Natural History Press.

Borchardt, Klaus-Dieter. 1987. *European Unification: The Origins and Growth of the European Community.* Luxembourg: Office for Official Publications of the European Communities.

Borneman, John. 1992. *Belonging in the Two Berlins: Kin, State, Nation.* Cambridge: Cambridge University Press.

Bourdieu, Pierre. 1975. "The Specificity of the Scientific Field and the Social Conditions of the Progress of Reason." *Social Science Information* 14 (6): 19–47.

———. 1977. *Outline of a Theory of Practice*. Translated by Richard Nice. Cambridge: Cambridge University Press.

———. 1984. *Distinction: A Social Critique of the Judgement of Taste*. Translated by Richard Nice. Cambridge: Harvard University Press.

———. 1988. *Homo Academicus*. Translated by Peter Collier. Stanford, CA: Stanford University Press.

Brenneis, Donald Lawrence. 1984. "Straight Talk and Sweet Talk: Political Discourse in an Occasionally Egalitarian Community." In *Dangerous Words: Language and Politics in the Pacific*, edited by D. L. Brenneis and F. Myers. New York: New York University Press.

———. 1994. "Discourse and Discipline at the National Research Council: A Bureaucratic *Bildungsroman*." *Cultural Anthropology* 9 (1): 23–36.

Brenneis, Donald Lawrence, and Fred R. Myers, eds. 1984. *Dangerous Words: Language and Politics in the Pacific*. New York: New York University Press.

Britan, Gerald M., and Ronald Cohen, eds. 1980. *Hierarchy and Society: Anthropological Perspectives on Bureaucracy*. Philadelphia: Institute for the Study of Human Issues.

Brucan, Silviu. 1988. "Europe in the Global Strategic Game." In *Europe: Dimensions of Peace*, edited by B. Hettne. London: Zed Books; Tokyo: United Nations University.

Buechler, Hans Christian, and Judith-Maria Buechler, eds. 1987. *Migrants in Europe: The Role of Family, Labor, and Politics*. New York: Greenwood Press.

Bull, Martin J. 1993. "Widening versus Deepening the European Community: The Political Dynamics of 1992 in Historical Perspective." In *Cultural Change and the New Europe: Perspectives on the European Community*, edited by T. Wilson and M. E. Smith. Boulder, CO: Westview Press.

Carrithers, Michael. 1989. "Sociality, Not Aggression, Is the Key Human Trait." In *Societies at Peace: Anthropological Perspectives*, edited by S. Howell and R. Willis. London: Routledge.

Certeau, Michel de. 1984. *The Practice of Everyday Life*. Translated by Steven F. Rendall. Berkeley and Los Angeles: University of California Press.

Clifford, James, and George E. Marcus, eds. 1986. *Writing Culture: The Poetics and Politics of Ethnography*. Berkeley and Los Angeles: University of California Press.

Cohen, Anthony P., ed. 1986. *Symbolising Boundaries: Identity and Diversity in British Cultures*. Manchester: Manchester University Press.

Collins, Randall. 1975. *Conflict Sociology: Toward an Explanatory Science*. New York: Academic Press.

Collins, Randall, and Sal Restivo. 1983a. "Development, Diversity, and Conflict in the Sociology of Science." *The Sociological Quarterly* 24 (2): 185–200.

———. 1983b. "Robber Barons and Politicians in Mathematics: A Conflict Model of Science." *The Canadian Journal of Sociology* 8 (2): 199–227.

Comaroff, Jean. 1985. *Body of Power, Spirit of Resistance: The Culture and History of a South African People*. Chicago: University of Chicago Press.

Comaroff, John L., and Simon Roberts. 1981. *Rules and Processes: The Cultural Logic of Dispute in an African Context*. Chicago: University of Chicago Press.

Coser, Lewis. 1968. "Conflict—Social Aspects." In *International Encyclopedia of the Social Sciences*, edited by D. Sills. 3:232–36.

Cozzens, Susan, and Thomas F. Gieryn, eds. 1990. *Theories of Science in Society*. Bloomington: Indiana University Press.

Crane, Diana. 1972. *Invisible Colleges: Diffusion of Knowledge in Scientific Communities*. Chicago: University of Chicago Press.

Dahl, Robert A. 1989. *Democracy and Its Critics*. New Haven: Yale University Press.

Derlega, Valerian J., and Janusz Grzelak, eds. 1982. *Cooperation and Helping Behavior: Theories and Research*. New York: Academic Press.

Dickson, David. 1987. "Space: It Is Expensive in the Major Leagues." *Science* 237 (4819): 1110–11. Special issue on "Science in Europe."

Douglas, Mary. 1966. *Purity and Danger: An Analysis of the Concepts of Pollution and Taboo*. London: Routledge and Kegan Paul.

Dubinskas, Frank A., ed. 1988. *Making Time: Ethnographies of High-Technology Organizations*. Philadelphia: Temple University Press.

Dumont, Louis. 1970. *Homo Hierarchicus: The Caste System and Its Implications*. Complete rev. English ed. Translated by Mark Sainsbury, Louis Dumont, and Basia Gulati. Chicago: University of Chicago Press.

———. 1986. *Essays on Individualism: Modern Ideology in Anthropological Perspective*. Chicago: University of Chicago Press.

Durkheim, Emile. 1915. *The Elementary Forms of the Religious Life*. Translated by Joseph Ward Swain. Reprint. 1965. New York: Free Press.

———. 1933. *The Division of Labor in Society*. Translated by George Simpson. Reprint. 1964. New York: Free Press.

Eisen, Arnold. 1987. "Called to Order: The Role of the Puritan *Berufsmensch* in Weberian Sociology." In *Max Weber, Rationality and Modernity*, edited by S. Whimster and S. Lash. London: Allen and Unwin.

Elshtain, Jean Bethke. 1981. *Public Man, Private Woman: Women in Social and Political Thought*. Princeton: Princeton University Press.

Epstein, A. L. 1978. *Ethos and Identity: Three Studies in Ethnicity*. London: Tavistock Publications.

Esman, Milton J., ed. 1977. *Ethnic Conflict in the Western World*. Ithaca, NY: Cornell University Press.

European Space Agency. 1983. *Europe into Space*. Paris: ESA Press and Publications Section.

———. 1985. *Forward to the Future*. Paris: ESA Press and Publications Section.

———. 1989. "Twenty-Five Years of European Cooperation in Space—Celebratory Ceremony on 19 April 1989." Special issue of the *ESA Bulletin*, no. 58: 7–39.

Evans-Pritchard, E. E. 1940. *The Nuer*. Oxford: Oxford University Press.

Fagan, Mary. 1988. "Peer Indicts UK's Arrogant Handling of Space Agency." *The Independent*, 5 October.

Flanagan, James G. 1989. "Hierarchy in Simple 'Egalitarian' Societies." *Annual Review of Anthropology* 18: 245–66.

Flanagan, James G., and Steve Rayner, eds. 1988. *Rules, Decisions, and Inequality in Egalitarian Societies*. Aldershot, England, and Brookfield, VT: Avebury.
Forsythe, Diana. 1993. "The Construction of Work in Artificial Intelligence." *Science, Technology, and Human Values* 18 (4): 460–79.
Fortes, Meyer. 1959. *Oedipus and Job in West African Religion*. Cambridge: Cambridge University Press.
Foucault, Michel. 1980. *Power/Knowledge: Selected Interviews and Other Writings 1972–1977*. Translated by Colin Gordon, Leo Marshall, John Mepham, and Kate Soper. Edited by C. Gordon. New York: Pantheon Books.
Frost, Peter J., and Larry F. Moore, Meryl Reis Louis, Craig C. Lundberg, and Joanne Martin. 1985. *Organizational Culture*. Beverly Hills, CA: Sage Publications.
Fuchs, Stephan. 1992. *The Professional Quest for Truth: A Social Theory of Science and Knowledge*. Albany: State University of New York Press.
Fujimura, Joan H. 1992. "Crafting Science: Standardized Packages, Boundary Objects, and 'Translation.'" In *Science as Practice and Culture*, edited by A. Pickering. Chicago: University of Chicago Press.
Galison, Peter, and B. Hevly, eds. 1992. *Big Science: The Growth of Large-Scale Research*. Stanford, CA: Stanford University Press.
Galtung, Johan. 1989. *Europe in the Making*. New York: Taylor and Fromais.
Geertz, Clifford. 1973. *The Interpretation of Cultures*. New York: Basic Books.
———. 1983. *Local Knowledge: Further Essays in Interpretive Anthropology*. New York: Basic Books.
Gerlach, Luther P., and Betty Radcliffe. 1979. "Can Independence Survive Interdependence?" *Futurics* 3 (3): 181–206.
Giddens, Anthony. 1971. *Capitalism and Modern Social Theory: An Analysis of the Writings of Marx, Durkheim and Max Weber*. Cambridge: Cambridge University Press.
———. 1984. *The Constitution of Society*. Berkeley and Los Angeles: University of California Press.
Gluckman, Max. 1956. *Custom and Conflict in Africa*. Oxford: Blackwell.
Greenhouse, Carol. 1986. *Praying for Justice: Faith, Order, and Community in an American Town*. Ithaca, NY: Cornell University Press.
———. 1992. "Signs of Quality: Individualism and Hierarchy in American Culture." *American Ethnologist* 19 (2): 233–54.
Greenwood, Davydd. 1984. *The Taming of Evolution: The Persistence of Nonevolutionary Views in the Study of Humans*. Ithaca, NY: Cornell University Press.
———. 1988. "Egalitarianism or Solidarity in Basque Industrial Cooperatives: The FAGOR Group of Mondragón." In *Rules, Decisions and Inequality in Egalitarian Societies*, edited by J. Flanagan and S. Rayner. Aldershot, England, and Brookfield, VT: Avebury.
Grillo, R. D., ed. 1980. *"Nation" and "State" in Europe: Anthropological Perspectives*. London: Academic Press.
Habermas, Jürgen. 1989a. "Social Action and Rationality." Reprinted in *Jürgen Habermas on Society and Politics: A Reader*, edited by S. Seidman. Boston: Beacon Press.

———. 1989b. "Technology and Science as Ideology." Reprinted in *Jürgen Habermas on Society and Politics: A Reader*, edited by S. Seidman. Boston: Beacon Press.

Hagendijk, Rob. 1990. "Structuration Theory, Constructivism, and Scientific Change." In *Theories of Science in Society*, edited by S. Cozzens and T. Gieryn. Bloomington: Indiana University Press.

Hagstrom, Warren O. 1965. *The Scientific Community*. New York: Basic Books.

Handelman, Don. 1981. "Introduction: The Idea of Bureaucratic Organization." *Social Analysis* 9: 5–23.

Handelman, Don, and Elliott Leyton. 1978. *Bureaucracy and World View: Studies in the Logic of Official Interpretation*. Social and Economic Studies, no. 22. St. John's: Institute of Social and Economic Research, Memorial University of Newfoundland.

Haraway, Donna. 1990. *Primate Visions: Gender, Race, and Nature in the World of Modern Science*. London: Routledge.

Harding, Sandra. 1986. *The Science Question in Feminism*. Ithaca, NY: Cornell University Press.

———. 1991. *Whose Science? Whose Knowledge? Thinking from Women's Lives*. Ithaca, NY: Cornell University Press.

Harding, Sandra, and Jean O'Barr, eds. 1987. *Sex and Scientific Inquiry*. Chicago: University of Chicago Press.

Herzfeld, Michael. 1987. *Anthropology through the Looking-Glass: Critical Ethnography in the Margins of Europe*. Cambridge: Cambridge University Press.

———. 1992. *The Social Production of Indifference: Exploring the Symbolic Roots of Western Bureaucracy*. Oxford: Berg.

Hess, David J. 1991. *Spirits and Scientists: Ideology, Spiritism, and Brazilian Culture*. University Park: Pennsylvania State University Press.

———. 1992. "Introduction: The New Ethnography and the Anthropology of Science and Technology." In *Knowledge and Society: The Anthropology of Science and Technology*, edited by D. Hess and L. Layne. Series edited by A. Rip. Vol. 9. Greenwood, CT: JAI Press.

Hess, David, and Linda Layne, eds. 1992. *Knowledge and Society: The Anthropology of Science and Technology*. Series edited by A. Rip. Vol. 9. Greenwich, CT: JAI Press.

Holmberg, David H. 1989. *Order in Paradox: Myth, Ritual, and Exchange among Nepal's Tamang*. Ithaca, NY: Cornell University Press.

Holmes, Douglas R. 1989. *Cultural Disenchantments: Worker Peasantries in Northeast Italy*. Princeton: Princeton University Press.

———. Forthcoming. *Radical Europe*. Princeton: Princeton University Press.

Horton, Robin, and Ruth Finnegan, eds. 1973. *Modes of Thought: Essays on Thinking in Western and Non-Western Societies*. London: Faber.

Howell, Signe. 1984. "Equality and Hierarchy in Chewong Classification." *JASO* 15 (1): 30–44.

Howell, Signe, and Roy Willis. 1989. "Introduction." In *Societies at Peace: Anthropological Perspectives*, edited by S. Howell and R. Willis. London: Routledge.

Hughes, Thomas P. 1987. "The Evolution of Large Technological Systems." In *The Social Construction of Technological Systems,* edited by W. Bijker, T. Hughes, and T. Pinch. Cambridge: MIT Press.

Hyde, Lewis. 1983. *The Gift: Imagination and the Erotic Life of Property.* New York: Vintage Books.

Kapferer, Bruce. 1988. *Legends of People, Myths of State: Violence, Intolerance, and Political Culture in Sri Lanka and Australia.* Washington, DC: Smithsonian Institution Press.

Keller, Evelyn Fox. 1985. *Reflections on Gender and Science.* New Haven: Yale University Press.

Knorr-Cetina, Karin. 1981. *The Manufacture of Knowledge: An Essay on the Constructivist and Contextual Nature of Science.* New York: Pergamon.

Krige, John. 1992. "The Prehistory of ESRO 1959/60." ESA HSR-1 (July). Noordwijk, the Netherlands: ESA Publications Division.

———. 1993. "Europe into Space: The Auger Years (1959–1967)." ESA HSR-8 (May). Noordwijk, the Netherlands: ESA Publications Division.

Lambright, W. Henry. 1994. "The Political Construction of Space Satellite Technology." *Science, Technology, and Human Values* 19 (1): 47–69.

Lane, Christel. 1981. *The Rites of Rulers: Ritual in Industrial Society: The Soviet Case.* Cambridge: Cambridge University Press.

Laqueur, Walter. 1982. *Europe since Hitler: The Rebirth of Europe.* Rev. ed. Harmondsworth, England: Penguin Books.

Latour, Bruno. 1987. *Science in Action: How to Follow Scientists and Engineers through Society.* Cambridge: Harvard University Press.

Latour, Bruno, and Steve Woolgar. 1979. *Laboratory Life: The Construction of Scientific Facts.* Rev. ed. Princeton: Princeton University Press.

Lederman, Rena. 1984. "Who Speaks Here? Formality and the Politics of Gender in Mendi, Highland Papua New Guinea." In *Dangerous Words: Language and Politics in the Pacific,* edited by D. L. Brenneis and F. Myers. New York: New York University Press.

Lévi-Strauss, Claude. 1963a. *Structural Anthropology.* Translated by Claire Jacobson and Brooke Grundfest Schoepf. New York: Basic Books.

———. 1963b. *Totemism.* Translated by Rodney Needham. Boston: Beacon Press.

Lloyd, Cathie, and Hazel Waters. 1991. "France: One Culture, One People?" *Race and Class* 32 (3): 49–65.

Longdon, Norman, ed. 1989. *ESA Annual Report '88.* Noordwijk, the Netherlands: ESA Publications Division.

Longdon, Norman, and Duc Guyenne, eds. 1984. *Twenty Years of European Cooperation in Space: An ESA Report.* Noordwijk, the Netherlands: ESA Scientific & Technical Publications Branch.

Lukes, Steven. 1982. Introduction to *The Rules of Sociological Method,* by Emile Durkheim. New York: Free Press.

Lüst, Reimar. 1987. *Europe and Space.* ESA Publication BR-35. Noordwijk, the Netherlands: ESA Publications Division.

———. 1989. "Foreword." *ESA Bulletin,* no. 58: 7.

Lutz, Catherine A. 1988. *Unnatural Emotions: Everyday Sentiments on a Microne-*

sian Atoll and Their Challenge to Western Theory. Chicago: University of Chicago Press.

Lynch, Michael. 1985. *Art and Artifact in Laboratory Science: A Study of Shop Work and Shop Talk in a Research Laboratory*. London: Routledge and Kegan Paul.

McDonald, Maryon. 1989. *'We Are Not French!'—Language, Culture and Identity in Brittany*. London: Routledge.

———. 1993. "The Construction of Difference: An Anthropological Approach to Stereotypes." In *Inside European Identities: Ethnography in Western Europe*, edited by S. Macdonald. Oxford: Berg.

Macdonald, Sharon, ed. 1993. *Inside European Identities: Ethnography in Western Europe*. Oxford: Berg.

McDonogh, Gary. 1993. "The Face behind the Door: European Integration, Immigration, and Identity." In *Cultural Change and the New Europe: Perspectives on the European Community*, edited by T. Wilson and M. E. Smith. Boulder, CO: Westview Press.

MacKenzie, Donald. 1987. "Missile Accuracy: A Case Study in the Social Processes of Technological Change." In *The Social Construction of Technological Systems*, edited by W. Bijker, T. Hughes, and T. Pinch. Cambridge: MIT Press.

Manno, V. 1988. "Competitive Selection of the Agency's Next Scientific Project." *ESA Bulletin*, no. 55: 8–9.

Marcus, George E., and Michael M. J. Fischer. 1986. *Anthropology as Cultural Critique: An Experimental Moment in the Human Sciences*. Chicago: University of Chicago Press.

Martin, Emily. 1987. *The Woman in the Body*. Boston: Beacon.

Marx, Karl. [1859] 1977. *Capital: A Critique of Political Economy*. Vol. 1. Translated by Ben Fowkes. New York: Vintage Books.

Mead, Margaret. [1937] 1961. *Cooperation and Competition among Primitive Peoples*. Enl. ed. with a new Preface and Appraisal. Boston: Beacon Press.

Merchant, Carolyn. 1980. *The Death of Nature*. New York: Harper and Row.

Merton, Robert K. 1968. *Social Theory and Social Structure*. Enl. ed. New York: Free Press.

———. 1973. *Sociology of Science*. Edited by B. Barnes. Harmondsworth, England: Penguin Books.

Mitroff, Ian. 1974. "Norms and Counter-Norms in a Select Group of the Apollo Moon Scientists." *American Sociological Review* 39: 579–95.

Moore, Sally Falk. 1978. *The Law as Process: An Anthropological Approach*. London: Routledge and Kegan Paul.

Morgan, Gareth. 1986. *Images of Organization*. Beverly Hills, CA: Sage Publications.

Mulkay, Michael. 1975. "Norms and Ideology in Science." *Social Science Information* 15: 637–56.

Myers, Fred R. 1986. *Pintupi Country, Pintupi Self: Sentiment, Place, and Politics among Western Desert Aborigines*. Washington, DC: Smithsonian Institution Press.

Myers, Fred R., and Donald Lawrence Brenneis, eds. 1984. "Introduction: Language and Politics in the Pacific." In *Dangerous Words: Language and Politics in the Pacific*, edited by D. L. Brenneis and F. Myers. New York: New York University Press.

Nader, Laura. 1968. "Conflict—Anthropological Aspects." In *International Encyclopedia of the Social Sciences*, edited by D. Sills. 3:236–41.

———. 1990. *Harmony Ideology: Justice and Control in a Zapotec Mountain Village*. Stanford, CA: Stanford University Press.

Nader, Laura, and Harry F. Todd, Jr., eds. 1978. *The Disputing Process: Law in Ten Societies*. New York: Columbia University Press.

Nisbet, Robert A. 1968. "Cooperation." In *International Encyclopedia of the Social Sciences*, edited by D. Sills. 3:384–90.

Noble, David F. 1992. *A World without Women: The Christian Clerical Culture of Western Science*. New York: Alfred A. Knopf.

Ortner, Sherry B. 1974. "Is Female to Male as Nature Is to Culture?" In *Women, Culture and Society*, edited by M. Z. Rosaldo and L. Lamphere. Stanford, CA: Stanford University Press.

———. 1984. "Theory in Anthropology since the Sixties." *Comparative Studies in Society and History* 26 (1): 126–66.

Ortner, Sherry B., and Harriet Whitehead, eds. 1981. *Sexual Meanings: The Cultural Construction of Gender and Sexuality*. Cambridge: Cambridge University Press.

Page, D. E., A. Pedersen, B. G. Taylor, and K. P. Wenzel, comps. 1986. *Report on the Activities of the Space Science Department in 1984–85*. ESA Publication SP-1082. Paris: ESA Publications Division.

Parsons, Talcott. 1947. Introduction to *The Theory of Social and Economic Organization*, by Max Weber. Translated by A. M. Henderson and Talcott Parsons. Edited by T. Parsons. Glencoe, IL: Free Press.

———. [1937] 1949. *The Structure of Social Action: A Study in Social Theory with Special Reference to a Group of Recent European Writers*. New York: Free Press.

Piaget, Jean. 1970. *Structuralism*. Translated by Chaninah Maschler. New York: Harper and Row.

Pieterse, Jan Nederveen. 1991. "Fictions of Europe." *Race and Class* 32 (3): 3–10.

Pinch, Trevor, and Wiebe Bijker. 1987. "The Social Construction of Facts and Artifacts: Or How the Sociology of Science and the Sociology of Technology Might Benefit Each Other." In *The Social Construction of Technological Systems*, edited by W. Bijker, T. Hughes, and T. Pinch. Cambridge: MIT Press.

Pondy, Louis R., Peter J. Frost, Gareth Morgan, and Thomas C. Dandridge, eds. 1983. *Organizational Symbolism*. Greenwich, CT: JAI Press.

Popper, Karl. 1961. *The Logic of Scientific Discovery*. New York: Basic Books.

Price, Derek J. de Solla. [1963] 1986. *Little Science, Big Science . . . and Beyond*. New York: Columbia University Press.

Provine, Marie. 1993. "The Human Rights of Women: A Feminist Assessment." Working Paper no. 2, Global Legal Studies Workshop. Bloomington: Indiana University School of Law.

Rabbie, Jacob M. 1982. "The Effects of Intergroup Competition and Cooperation on Intragroup and Intergroup Relationships." In *Cooperation and Helping Behavior: Theories and Research*, edited by V. Derlega and J. Grzelak. New York: Academic Press.

Radcliffe-Brown, A. R. 1952. *Structure and Function in Primitive Society*. Glencoe, IL: Free Press.

Rapp, Rayna. 1991. "Moral Pioneers: Women, Men, and Fetuses on a Frontier of Reproductive Technology." In *Gender at the Crossroads of Knowledge*, edited by M. di Leonardo. Berkeley and Los Angeles: University of California Press.

Räthzel, Nora. 1991. "Germany: One Race, One Nation?" *Race and Class* 32 (3): 31–48.

Restivo, Sal. 1988. "Modern Science as a Social Problem." *Social Problems* 35 (3): 206–25.

Ries, Nancy V. 1993. "Mystical Poverty and the Rewards of Loss: Russian Culture and Conversation during Perestroika." Ph.D. diss., Cornell University.

Rossiter, Margaret. 1982. *Women Scientists in America: Struggles and Strategies to 1840*. Baltimore, MD: Johns Hopkins University Press.

Russo, Arturo. 1993a. "The Definition of a Scientific Policy: ESRO's Satellite Programme in 1969–1973." ESA HSR-6 (March). Noordwijk, the Netherlands: ESA Publications Division.

———, ed. 1993b. *Science beyond the Atmosphere: The History of Space Research in Europe*. Proceedings of a symposium held in Palermo, 5–7 November 1992. ESA HSR-Special (July). Noordwijk, the Netherlands: ESA Publications Division.

Sahlins, Peter. 1989. *Boundaries: The Making of France and Spain in the Pyrenees*. Berkeley and Los Angeles: University of California Press.

Scarselletta, Mario. 1993. "Button Pushers and Ribbon Cutters: Observations on Skill and Practice in a Hospital Laboratory and Their Implications for the Shortage of Skilled Technicians." EQW Working Paper no. WP12. Philadelphia: National Center on the Educational Quality of the Workforce, University of Pennsylvania.

Schama, Simon. 1987. *The Embarrassment of Riches: An Interpretation of Dutch Culture in the Golden Age*. New York: Alfred A. Knopf.

Schiebinger, Londa. 1989. *The Mind Has No Sex? Women in the Origins of Modern Science*. Cambridge: Harvard University Press.

Schwartzman, Helen B. 1989. *The Meeting: Gatherings in Organizations and Communities*. New York: Plenum Press.

Segal, Daniel A. 1991. "'The European': Allegories of Racial Purity." *Anthropology Today* 7 (5): 7–9.

Shapin, Steve. 1989. "The Invisible Technician." *American Scientist* 77 (Nov.–Dec.): 554–63.

Simmel, Georg. [1908] 1955. "Conflict." In *Conflict and the Web of Group-Affiliations*. Translated and edited by Kurt H. Wolff. Glencoe, IL: Free Press.

Sluka, Jeffrey A. 1992. "The Anthropology of Conflict." In *The Paths to Domination, Resistance, and Terror*, edited by C. Nordstrom and J. Martin. Berkeley and Los Angeles: University of California Press.

Stephens, Sharon. 1993. "The Making of an Invisible Event: Assessing Risks and Negotiating Identities in Post-Chernobyl Norway." Paper presented at the Annual Meeting of the American Anthropological Association, Washington, DC.

Strathern, Marilyn, ed. 1987. *Dealing with Inequality: Analysing Gender Relations in Melanesia and Beyond*. Cambridge: Cambridge University Press.

Tambiah, Stanley Jeyaraja. 1990. *Magic, Science, Religion, and the Scope of Rationality*. Lewis Henry Morgan Lecture Series. Cambridge: Cambridge University Press.

Taussig, Michael. 1980. *The Devil and Commodity Fetishism in South America*. Chapel Hill: University of North Carolina Press.

———. 1987. *Shamanism, Colonialism, and the Wild Man: A Study in Terror and Healing*. Chicago: University of Chicago Press.

Taylor, Charles. 1989. *Sources of the Self: The Making of the Modern Identity*. Cambridge: Harvard University Press.

———. 1992. *The Ethics of Authenticity*. Cambridge: Harvard University Press.

Tonkin, Elizabeth, Maryon McDonald, and Malcolm Chapman, eds. 1989. *History and Ethnicity*. London: Routledge.

Townshend, John. 1980. "European Cultural Co-operation." In *European Cooperation Today*, edited by K. Twitchett. London: Europa Publications.

Traweek, Sharon. 1988. *Beamtimes and Lifetimes: The World of High Energy Physicists*. Cambridge: Harvard University Press.

———. 1992. "Border Crossings: Narrative Strategies in Science Studies and among Physicists in Tsukuba Science City, Japan." In *Science as Practice and Culture*, edited by A. Pickering. Chicago: University of Chicago Press.

Trevarthen, Colwyn, and Katerina Logotheti. 1989. "Child in Society, and Society in Children: The Nature of Basic Trust." In *Societies at Peace: Anthropological Perspectives*, edited by S. Howell and R. Willis. London: Routledge.

Turner, Terrence. 1977. "Transformation, Hierarchy and Transcendence: A Reformulation of Van Gennep's Model of the Structure of Rites De Passage." In *Secular Ritual*, edited by S. F. Moore and B. Myerhoff. Amsterdam: Van Gorcum Press.

Turner, Victor. 1967. *The Forest of Symbols: Aspects of Ndembu Ritual*. Ithaca, NY: Cornell University Press.

———. 1968. *The Drums of Affliction: A Study of Religious Processes among the Ndembu of Zambia*. Symbol, Myth, and Ritual Series. Ithaca, NY: Cornell University Press.

———. 1974. *Dramas, Fields, and Metaphors*. Ithaca, NY: Cornell University Press.

Twitchett, Carol Cosgrove. 1980. "The EEC and European Co-operation." In *European Cooperation Today*, edited by K. Twitchett. London: Europa Publications.

Twitchett, Kenneth J. 1980. "European Regionalism in Perspective." In *European Cooperation Today*, edited by K. Twitchett. London: Europa Publications.

Varenne, Hervé. 1977. *Americans Together: Structured Diversity in a Midwestern Town*. New York: Columbia University Press.

———. 1993. "The Question of European Nationalism." In *Cultural Change and the New Europe: Perspectives on the European Community*, edited by T. Wilson and M. E. Smith. Boulder, CO: Westview Press.

Webber, Frances. 1991. "From Ethnocentrism to Euro-racism." *Race and Class* 32 (3): 11–17.

Weber, Max. 1946. "Bureaucracy." In *From Max Weber: Essays in Sociology.* Translated, edited, and with an Introduction by H. H. Gerth and C. Wright Mills. New York: Oxford University Press.

———. 1947. *The Theory of Social and Economic Organization.* Translated by A. M. Henderson and Talcott Parsons. Edited and with an Introduction by T. Parsons. Glencoe, IL: Free Press.

———. 1958. *The Protestant Ethic and the Spirit of Capitalism.* Translated by Talcott Parsons. New York: Charles Scribner's Sons.

Weiner, Annette. 1976. *Women of Value, Men of Renown: New Perspectives in Trobriand Exchange.* Austin: University of Texas Press.

———. 1992. *Inalienable Possessions: The Paradox of Keeping-While-Giving.* Berkeley and Los Angeles: University of California Press.

Wilson, Thomas M. 1993. "An Anthropology of the European Community." In *Cultural Change and the New Europe: Perspectives on the European Community,* edited by T. Wilson and M. E. Smith. Boulder, CO: Westview Press.

Wilson, Thomas M., and M. Estellie Smith, eds. 1993. *Cultural Change and the New Europe: Perspectives on the European Community.* Boulder, CO: Westview Press.

Zabusky, Stacia E. n.d. "Aspiration and Ideology in the Construction of Community: American Scientists Defend Utopia 1978–1985." Manuscript.

———. 1992. "Multiple Contexts, Multiple Meanings: Scientists in the European Space Agency." In *Knowledge and Society: The Anthropology of Science and Technology,* edited by D. Hess and L. Layne. Series edited by A. Rip. Vol. 9. Greenwood, CT: JAI Press.

———. 1993a. "Expressions of Diversity: 'Nationality' in Discourses of Work and Leisure in a European Context." Paper presented in the Department of Anthropology Colloquium Series, Cornell University.

———. 1993b. "Multinational or Supranational?—Boundaries and Identities in a European Organization." Paper presented in the Institute for European Studies Seminar Series, Cornell University.

———. 1993c. "Strain and the Organizational Scientist: A Cultural Explanation." EQW Working Paper no. WP16. Philadelphia: National Center on the Educational Quality of the Workforce, University of Pennsylvania.

Index

Ingram Content Group UK Ltd.
Milton Keynes UK
UKHW012014090623
423193UK00001B/117

9 780691 029726